Molecules of Murder
Criminal Molecules and Classic Cases

Molecules of Murder
Criminal Molecules and Classic Cases

John Emsley

RSCPublishing

ISBN: 978-0-85404-965-3

A catalogue record for this book is available from the British Library

Published by The Royal Society of Chemistry,
Thomas Graham House, Science Park, Milton Road,
Cambridge CB4 0WF, UK

Registered Charity Number 207890

For further information see our web site at www.rsc.org

Preface

The crime of murder by deliberate poisoning has always held a terrible fascination, not least because of the secrecy surrounding its planning and the difficulty inherent in its detection. In *Molecules of Murder* I have viewed a range of examples of such crimes, not through the eyes of the criminal, the detective, nor even the victim, but from the point of view of the forensic scientist, and with special focus on the toxic agents involved. We will be looking at ten poisons used in famous murders, five of which were natural chemicals (Part I) and five of them produced by humans (Part II). Each chapter starts with a discussion of the poison itself and ends with a detailed examination of one or more murders in which it has been used.

Popular science may have its following, but true crime has an even greater appeal. *Molecules of Murder* is intended for both groups of reader. I hope that those who like true crime will enjoy reading about murder by poisoning from a slightly different angle. I also hope that those who are interested in science and its role in life—and of course death—will find the toxicology interesting, as well as the murder cases in which it has played a key part.

The book is dedicated to those unsung heroes, the analytical chemists, the men and women who are able to find the proverbial needle in the haystack, and who indeed can sometimes find it in a haystack the size of the Great Pyramid. Nowhere are their skills more valuable to society than in solving crime, and while murder by poison may be relatively rare, they are the ones who can supply the vital answers. They play a key role in the complex struggle

Molecules of Murder: Criminal Molecules and Classic Cases
By John Emsley
© John Emsley, 2008

against all kinds of substance abuse, and they now have their own International Association of Forensic Toxicologists, founded in London in 1963.

This book is about a few highly dangerous chemicals. Sadly, some people think that *all* chemicals are dangerous. Their judgement is of course coloured by the way the word *chemical* is now used by many to imply something intrinsically harmful. Also misguided are those who consider there is a distinction to be made between natural and man-made chemicals. They assume that natural means safe, and man-made means hazardous. This group of people may find that the first part of this book comes as a surprise. If you are such a person, please keep an open mind. I hope you will soon realise that across the spectrum of toxic chemicals, those which Nature produces can be at least as deadly as any devised by humans.

That people are increasingly interested in science is shown by the popularity of articles, books and television programmes on subjects as diverse as astronomy and zoology. For some reason chemistry still features less than most scientific topics in terms of popular appeal, but its forensic role has given it a higher profile in recent years. Even so, those programmes which embrace forensic science usually emphasise the work of pathologists and rarely provide details of forensic chemistry. Naturally a scene set in a chemical laboratory will always be less dramatic than one with naked bodies and dissected cadavers. When the story is told in a book, however, the lurid fascination of the mortuary is no longer present and the chemical laboratory takes centre stage.

I hope *Molecules of Murder* will make you want to learn more about the area of science I find most fascinating. People and their darker doings are always going to be more interesting than chemists in white coats. However, by bringing the two together I hope I can both inform and entertain, and encourage some appreciation of the roles of pharmaceutical, forensic and research chemists. Maybe if you are young it might even encourage you to join our ranks.

Whether we are aware of it or not, chemistry touches virtually every aspect of our lives. More than any other science it has transformed the world in which we live, bringing us abundant food, better health, stronger materials, softer fabrics, brighter colours, cleaner homes, safer transport, greater energy efficiency, and more advanced plastics. But chemistry also has its darker side, and that has been in discovering and developing substances which are potentially harmful. Toxicology is the science which studies all such materials, which only become a threat to humans when they are misused.

Nowadays it is almost impossible for the ordinary person to obtain toxic chemicals—and it is almost impossible for a poisoner to escape detection when they are used. The result has been to make murder by poison, so beloved of the mystery writers of old, a rare event in real life.[i] Poison is the weapon chosen by those who wish to kill but do so in such a way that the victim is unaware of it, and who want the death to be assumed natural. When such murderers achieve their aim they may secretly celebrate. But when they fail, and they are brought to trial and found guilty, then we can all celebrate. However, as this book shows, some poisoners are never found, and others are found and then found not guilty, which makes re-analysing these cases all the more fascinating.

Finally, a warning to readers who are of a very serious frame of mind. I do on occasions make asides that you might find rather too jocular for a book about poisons. Let me apologise in advance for these remarks, but I am of the opinion that what humans do is to a certain extent for the amusement of the gods and believe we should view it accordingly.

---

[i] For example, in 1989 there were 19 000 murders in the USA of which fewer than 30 were caused by poison, although no doubt there are a number which remain undetected.

Contents

Molecules of Murder: Criminal Molecules and Classic Cases
By John Emsley
© John Emsley, 2008

Acknowledgements

I should like to thank the following people for their help in writing this book. Some of them provided me with information I would not otherwise have had, some checked that the contents of chapters were scientifically correct, and some were even prevailed upon to read the completed manuscript. In alphabetical order they are as follows:

- Dr Rob Allott of the UK Environment Agency, who checked Chapter 10.
- Sr Ann-Marie FJ MA, of the Rectory, Charmouth, Dorset, who explained why Georgi Markov is buried in the graveyard of the parish church at Whitchurch Canonicorum.
- Dr Alan Bailey of the Analytical Services Centre of the Forensic Science Service, London, who vetted Chapters 1, 3, 9, 10, and the Glossary.
- Katy Bridgwood, a University of Cambridge research chemist, drew the molecular structures shown in the Glossary.
- Dr Catherine J Duckett, Lecturer in Analytical and Forensic Chemistry at Keele University, appraised Chapters 4, 6 and 8 and came up with useful tips for improving them.
- My wife, Joan Emsley, read the complete text and pointed out those parts where it appeared I was assuming too much chemical knowledge on the part of the general reader, for whom this book has also been written.
- Mrs Christine Feather of Stirton, near Skipton, read the chapter on heroin and Harold Shipman and made useful comments.
- Dr Bernard Hart of Congleton, Cheshire, and former medical officer for ICI Plant Protection Division, checked Chapter 9 on paraquat.

- Professor Brian Mann of Sheffield University scrutinised Chapter 7 on carbon monoxide and made useful comments.
- Dr Gillian Marsh, a toxicologist of Northumberland, reviewed the Introduction as well as Chapters 1, 3, 5 and the Glossary.
- Professor Steve Ley and Rose Ley, of the Department of Chemistry, University of Cambridge, read the complete text and made invaluable suggestions for improving it.
- Dr Lewis Smith, a toxicologist working for Syngenta, reviewed Chapter 9.
- Dr Michael Utidjian of Wayne, New Jersey, was particularly helpful with advice about the effects of various toxins and their histories. Michael served as the Corporate Medical Director of the former American Cyanamid Co, at one time a major producer of cyanide.

Things to Bear in Mind

Molecules of Murder is a popular science book and as such may use terms that you are not familiar with. Where a technical word in the text is shown in bold then there is more information given in the Glossary. Because I hope this book will be read by those who are not chemists, I use the name by which a chemical is best known. When talking about specific chemicals the Glossary will provide the correct chemical name and its molecular structure.

Units of measurement: Very little of a poison may be needed to produce a toxic response or be present in a victim's body after death, and special units are required in order to discuss the tiny amounts detected by analytical chemists. These units are milligrams (mg), which are a thousandth part of a gram, and micrograms (µg), which are a millionth part of a gram. Other ways of expressing small amounts are parts per million (ppm) or parts per billion (ppb). The Glossary gives further details regarding **units**.

Currencies change in value from year to year, and even from day to day, in relation to other currencies. I have tried to give some idea of what things in the past would now cost, assuming that inflation has decreased the value of the pound sterling by a factor of 100 since the time of Queen Victoria, so that what £1 would buy then would cost around £100 today. Comparing prices across different currencies is easier, but even these can change significantly within a matter of months. At the time of writing, the UK pound (£) is worth around two US dollars ($), and this conversion factor has been used throughout.

Milestones in the Forensic Analysis of Poisons

Before chemistry began to develop out of alchemy in the late 1600s it was impossible to *prove* that a death had occurred through poisoning, whereas if there is the least suspicion today that someone has been poisoned it will be confirmed. Forensic chemists have an armoury of tests and equipment which can measure the quantity of substances present in blood or tissue, even when this is as little as a trillionth of a gram, which is so tiny it would be invisible to the human eye. There is a story of one forensic chemist who boasted he knew of a poison that no one could detect, and to prove his point he committed suicide with it—but he was wrong. He had poisoned himself with **fluoroacetate**. He knew that only a minute dose was needed to kill and that it would disperse itself into every cell of the body. There it would block the essential enzymes which normally process acetate ions, a chemical of which the body produces a great deal. Who then would be able to trace a few fluoroacetate ions among the sea of normal acetate ions? The answer turned out to be his forensic colleagues, who rose to the challenge and were eventually able to prove what he had used.

One of the earliest forensic investigations of a poison was carried out before the trial of an aristocratic lady, the Marquise de Brinvilliers (1630–1676). She lived near Paris and was suspected of poisoning her father and brothers in order to inherit their wealth. Her lover provided the poison, and all was revealed when he suddenly died in 1672 and a locked cabinet in his room was opened. Inside were her incriminating letters and a white powder which was suspected of being the poison he had supplied her with. An apothecary, Guy Simon, was asked to analyse this and he carried out a few primitive chemical observations. He noted that it was soluble in water, and that the solution obtained did not form a precipitate with either oil of tartar, which was a concentrated solution of potassium carbonate, or with sea water, which is essentially a solution of sodium chloride. When Simon heated the powder he observed that it sublimed, in other words it evaporated in the solid state without decomposition, and was deposited as white crystals around the neck of the vessel. When it was fed to a pigeon, a cat and a dog, they all died. The court accepted his evidence that the powder was indeed the infamous white **arsenic** (As_2O_3). The Marquise de Brinvilliers was tried, found guilty, tortured until she confessed her guilt, and finally beheaded in 1676.

A hundred years later chemistry was still in its infancy, but it was now possible for white arsenic to be more conclusively identified. The Dutch physician, Hermann Boerhaave (1668–1738), suggested that this could be achieved by placing a sample of the poison on hot charcoal and noting the garlic-like odour it gave off. A judge and jury needed more positive evidence than this when it came to identifying the poison which Mary Blandy (1720–1752) had used to kill her father. The Blandys lived at Henley-on-Thames in Oxfordshire, and she poisoned her father with a white powder sent by her lover, Lieutenant William Cranstoun, who assured her it would end her father's objection to their marriage. It did—it killed him. It also made a servant very ill. At Mary's trial, a Dr Addington and a Mr King provided the forensic evidence that the powder was in fact white arsenic, and they did so by comparing its properties with those of a sample of the real thing. They noted that each gave off the same smell when heated, and each had similar solubility in water; when other solutions were added to the dissolved material the same coloured precipitates were formed. The court was convinced that what Mary had put in her father's food was indeed white arsenic. She was convicted of murder, and publicly hanged on 6 April 1752. (Her lover fled to France and died of a fever later that year.)

As chemistry developed in the late 1700s so did the techniques of chemical analysis. Around 1790 a chemist, Johann Metzger, noticed that if samples of tissues and fluids thought to be contaminated with arsenic were heated strongly, they gave off a vapour which condensed on a cold plate as white crystals. In 1806, Valentine Rose of the Berlin Medical Faculty showed that if the stomach contents of someone who had died of arsenic poisoning were subjected to Metzger's test, the arsenic could be identified. The science of poisons was taken to a higher level by Mathieu Joseph Bonaventure Orfila (1787–1853). He was a professor of legal medicine and chemistry in the University of Paris, and in 1814, when he was only 26, he published the first textbook on the subject, *Traité de Toxicologie Générale (A Treatise of General Toxicology)*. In it he divided poisons into the following categories: corrosives, astringents, acrids, narcotics and putrefacients. It established his international reputation, and he is now looked upon as the founder of this branch of knowledge.

In cases of murder by poisoning, the one most commonly used was white arsenic. Proving that this was the cause of death became the single most important forensic challenge, a challenge which was taken up by one chemist whose evidence in a murder case had been mocked in court. He eventually became one of the most famous names in the history of forensic investigations—James Marsh (1794–1846). He had been asked to identify the poison which in 1832 had killed the 80-year-old grandfather of John Bodle, who stood to inherit his estate worth £20 000 (equivalent to £2 million today). The death was suspicious and Marsh, who worked as an ordnance chemist at the Royal Arsenal at Woolwich, London, was asked to test the suspect powder and show that arsenic was present in the organs taken from the dead man. He did this by precipitating the arsenic as the bright yellow sulfide, but when his samples were submitted in evidence they had become discoloured. Bodle's defence counsel was able to convince the jury that they were therefore unreliable and Bodle was found not guilty. He later confessed to the murder, however, in the full knowledge that under English law a person could not at that time be tried twice for the same crime. (This law no longer applies.)

Marsh was upset by his failure to provide the solid evidence that could have convicted Bodle, and he set about searching for a forensic test which would in future convince a jury that arsenic was really present. His new method was to heat a sample of the suspect tissue with concentrated hydrochloric acid until it had completely dissolved. A piece of zinc was then added to the solution, and if arsenic was present it generated **arsine** gas (AsH_3). When this gas was passed along a heated glass tube it decomposed to arsenic, which was deposited as a shiny mirror. If the tube was then sealed, the arsenic remained unchanged, and it could be shown to the court and passed around by members of the jury. He published his new method in 1836 and it was regarded as unquestioned analytical proof in criminal trials. Jurors were invariably impressed, and it became the standard test for arsenic poisoning for more than 100 years.

In 1842 Edgar Reinsch devised an alternative test for arsenic that could be conducted at the crime scene. A suspect sample was heated in a test tube with concentrated hydrochloric acid until dissolved and a copper strip was dipped into the solution. Arsenic would deposit on the copper, although this could take up to an hour, but it was possible to detect a tenth of a microgram by this method.

As the 1800s progressed the science of chemistry rapidly developed, and by the end of the century a wide range of poisons could be identified in the laboratory. One of the most easily detected non-metallic poisons was white **phosphorus**, which was used for making matches and to kill vermin—and sometimes to murder humans. It is highly toxic, but it could be revealed if a suspect sample was heated in a totally blacked-out room. Then it would phosphoresce with an eerie yellow-green glow as it vaporised. When the vapour was condensed and collected in water it could be oxidised to phosphate, and then converted to insoluble yellow ammonium phosphomolybdate, $(NH_4)_3PO_4.12MoO_3$. The use of this compound effectively multiplied the weight of the phosphorus present by a factor of 57, so that 1 ppm of phosphorus in a litre of the solution gave a weighable 57 mg of precipitate.[i]

Metallic poisons were much easier to test for than those of organic origin, since it was relatively easy to convert them to insoluble metal compounds which could be precipitated, filtered off and weighed. Organic poisons gathered from plants were much more difficult to detect, and they had been known for thousands of years. These presented a considerable challenge for forensic chemists, but methods were devised in the 1800s to isolate the active ingredients, purify them and then characterise them in some way, perhaps by their melting point if they were crystalline. In this manner morphine was extracted from opium in 1805, strychnine from *nux vomica* in 1819, coniine from hemlock in 1826, nicotine

---

[i]The comparative weights of ammonium phosphomolybdate and phosphorus are 1842 and 32.

from tobacco in 1828, and atropine from belladonna in 1833. Extracting them from the organs of a dead body was much more difficult, however, and indeed proved impractical during these early years. Even the famous toxicologist, Orfila, doubted that it would ever be possible.

Nevertheless, in 1850 the Belgian chemist, Jean Servais Stas, developed a method of extracting **nicotine** from the organs of a murder victim, and this led to the conviction of Count de Bocarme. The Count had poisoned his wife's ailing brother, Gustave Fougnies, with nicotine so that they could inherit his wealth, but that was in jeopardy once Gustave announced his intention of marrying. The Count invited him to visit them, whereupon he died. The doctor who examined the body thought at first that the cause of death was a stroke, but the condition of the throat and stomach suggested that these were inflamed by some kind of irritant. This was actually a solution of nicotine in wine. Stas was called to investigate and realised that he needed to extract nicotine from the samples taken from the body. He eventually discovered that the solvent **diethyl ether** was able to do this. Filtration and evaporation of the extract yielded enough material for it to be identified as nicotine. The Count was found guilty and executed.

Chemical techniques became progressively more sophisticated, until by the beginning of the twentieth century it was possible to extract evidence from even the most unlikely samples. The forensic investigation of the few fleshy remains of Dr Crippen's wife in 1910 was a triumph of forensic analysis due to the minute traces of the poison hyoscine present (Chapter 2). One of the UK's greatest forensic scientists was Sir Bernard Spilsbury (1877–1947), who had been involved in the Crippen case and became the country's first celebrity forensic scientist. Such was his pre-eminence in the profession and the absolute certainty with which he presented his evidence that few dared to question his findings, although there have since been suspicions that he was occasionally responsible for a miscarriage of justice. We will meet Spilsbury again in Chapter 7, which is concerned with the poison carbon monoxide, because he sadly committed suicide with this deadly gas in his laboratory at University College, London, in 1947. (He was depressed following a series of strokes.)

From the viewpoint of forensic toxicology, substances may be classified according to the methods needed to isolate them from biological materials, rather than by their chemical properties. The classes are:

- gaseous or volatile substances, which are isolated by diffusion or distillation;
- metals, which can be analysed following removal of the biological material by means of a strong oxidising acid;
- toxic non-metals in the form of negative ions, which can be isolated by dialysis or ion-exchange methods; and
- non-volatile organic compounds, which are isolated by liquid–liquid or liquid–solid extraction processes. This last category is sub-classified according to whether the toxin can be extracted under acid, alkaline or neutral conditions.

The introduction of paper **chromatography** in 1906 permitted more sophisticated levels of separation. Today, much more powerful types of chromatography are available, such as thin-layer chromatography (TLC), high performance thin-layer chromatography (HPLC), gas chromatography linked to **mass spectrometry** (GCMS), and liquid chromatography connected to double mass spectrometric analysis (LCMSMS). Around the middle of the twentieth century forensic chemists began to use spectroscopic techniques to identify poisons in minute quantities, which was possible by analysing their ultraviolet, infrared or visible spectra. Using current GCMS methods there is no longer any need to extract large amounts of tissue using solvents—a few micrograms will suffice. It is now possible to identify even the tiniest amounts of a toxin in any organ of the body, including the vitreous humour of the eyeball.

In the early 1950s it even became possible to analyse a strand of hair for toxic elements, and this was how, in 1958, US Sergeant Marcus Marymont was proved to have poisoned his wife when he returned on leave to his family in Ruislip, northwest London. Her hair revealed bands of arsenic occurring at intervals corresponding to his visits. He had attempted a number of times to poison her, but had only succeeded in making her very ill. His motive for murder was to be free to marry his new girlfriend. The method employed to detect and measure the arsenic, and thereby to convict Marymont, was **neutron activation** analysis, a technique requiring access to a nuclear reactor.

Poisoners have always tended to assume that once a body is cremated all evidence of the poison will disappear. This is in fact true of organic poisons, as they are destroyed in the flames. Harold Shipman knew this, and always endeavoured to persuade his victims' relatives to have the remains cremated, and most of them did—but not all, and it was those few who had been buried who provided the evidence which convicted him (Chapter 4). On the other hand, one poisoner was partly convicted by the analysis of his victim's cremated body, due to the fact that burning does not destroy metallic atoms such as the thallium he had used to kill his workmates. His name was Graham Young and he was convicted of murder in 1972. (He died in prison in 1990.) The ashes of one of his victims, Bob Egle, were analysed and found to contain 9 mg of thallium.

When a death is suspected of being due to unnatural causes, there is a certain amount of tissue and body fluid which has to be collected. This includes blood from the heart (typically 25 ml), femoral blood (10 ml), liver (100 g) and kidney (50 g), together with all the bile, the entire quantity of urine in the bladder, the vitreous humour from the eyes, and the total contents of the stomach. Samples of brain tissue and the spleen may also be required. If any samples are found to contain a suspicious chemical then it may also be necessary to take specimens of the lungs and intestines. In the case of bodies which have partially decomposed, samples of muscle, skin, bone, hair, nails—and even the maggots on the body—are collected for analysis.

An excellent account of forensic toxicology is given in Chapter 12 of the book *Crime Scene to Court*[ii] by Robert Anderson. Poisons may be categorised as:

- corrosive, for example concentrated sulfuric acid or sodium hydroxide;
- irritant, such as the metals arsenic and mercury;
- systemic, which are those which disrupt the biochemical processes of living cells; and
- toxins, the natural chemical poisons produced by living matter ranging from pathogenic microbes to insects and snakes. Strictly speaking, the term toxin applies only to these last of these.

[ii] Edited by Peter White, published by the Royal Society of Chemistry in 2004 (2nd edition).

A forensic toxicologist will generally know which molecules need to be looked for. When the cause of death is suspicious but the poison is unknown, then the toxicologist must begin a complex and systematic analysis of a wide variety of samples for the presence of every kind of poison. The samples most likely to yield the answer are, in order of probability, blood, urine, liver, bile, vitreous humour, cerebrospinal fluid (from the brain), brain tissue, lungs, hair, and nail clippings. Urine is less useful than blood, since the amount present in the urine is highly dependent on what the victim had been drinking, the degree of hydration of the body, and also the condition of the kidneys, which often contain only metabolites of the suspected poison or drug.

Samples have to be pre-treated to make them suitable for extracting the poison. In the case of blood this involves centrifugation, urine requires filtration, and in the case of tissue it means homogenisation. It may also be necessary to adjust the degree of acidity. The poison is likely to be present in tiny amounts, which makes it better to extract it using a solvent which is immiscible with water, such as **diethyl ether** or **chloroform**. Alternatively, it might be possible to adsorb it on to a solid material such as **silica**. Having extracted the poison it has to be purified, the so-called clean-up stage. Then the identification of the poison can begin, using one or more spectrometric techniques.

The forensic toxicologist is generally concerned with analysing samples of blood and urine from living people, such as those suspected of having used recreational substances, consuming too much alcohol, or of taking a drug to boost their performance at sport. It is only rarely that they are called upon to analyse for poisons like those discussed in this book, which have been given in a deliberate attempt to kill. They may however analyse for carbon monoxide or hydrogen cyanide in blood samples taken from victims of fires, since both of these molecules can be generated in such cases. The hydrogen cyanide comes from the burning of textiles containing man-made polymers such as polyurethane or polyacrylonitrile, or from biopolymers such as wool or silk. Carbon monoxide can arise from the burning of any carbon-containing substance.

In developed societies the most common cause of fatal poisonings is legal drugs (*e.g.*, painkillers, used to commit suicide) or illegal drugs (*e.g.*, a heroin overdose), whereas in less developed territories the poisons causing death are more often pesticides, and in most cases through accidental misuse.

Murder by poison is now quite rare, and becoming rarer. Maybe one day it will be seen as a crime of previous ages, but until that happy time we can savour its forbidden fruits in those cases where it has been the wilful instrument of death.

Part I:
Some of Nature's Deadly Chemicals:
From Medicine to Murder

CHAPTER 1

Ricin and the Rolled Umbrella

A word shown in **bold** *indicates that further information will be found in the Glossary*

1.1 Waterloo Sunset

An umbrella is not the most obvious way of delivering a fatal dose of poison, but that was the weapon chosen by the Bulgarian secret service in 1978. Its targets were 26-year-old Vladimir Kostov and 49-year-old Georgi Markov, dissidents who had defected to the West and were embarrassing the corrupt communist dictatorship of President Todor Zhivkov by broadcasting on the US propaganda station Radio Free Europe. Kostov survived, but Markov died.

On 7 September 1978 Markov parked his car near Waterloo Station and was waiting for a bus to take him over the River Thames to Bush House, where he worked for the BBC World Service. Suddenly he felt a sharp pain in his leg, and turned to find a man behind him with an umbrella which he had dropped on to the pavement. Although Markov did not know it, the umbrella had fired a tiny metal pellet no bigger than a pin-head into his thigh. By the time Markov reached his home in Balham that evening he was already feeling very ill. He began vomiting and was running a high fever.

Markov was unaware that beneath the skin of his leg was a minute pellet containing one of the deadliest of all poisons. He died in agony four days later despite being in intensive care. The following week I was drawn into the story, when I was rung up by *The Guardian*

Molecules of Murder: Criminal Molecules and Classic Cases
By John Emsley
© John Emsley, 2008

newspaper to be told about the pellet that had killed him. Could I speculate about the type of toxic chemical which it might have contained? I guessed a spider or snake venom, but I was wrong. It was something far more deadly, and originating from a plant; it was almost certainly **ricin**.

1.2   Toxicology and Chemistry

Ricin is produced by the castor bean plant, *Ricinus communis*. It makes this toxic protein molecule in its seeds, presumably to deter predators. Once the seed has germinated the toxin is no longer needed and it disappears. Until this happens the ricin is indeed a very effective deterrent, deadlier even than the most powerful of the nerve gases. The toxicity of a poison can be judged by its LD_{50}. For ricin it is 0.1 micrograms per kilogram ($\mu g/kg$) of body weight, whereas the figure for the most lethal nerve gas known, coded VX, is as high as $20\,\mu g/kg$. Although these measurements were made using rodents, if they were to apply to a human they would suggest that for someone of average weight, say 70 kg, a dose of a mere $7\,\mu g$ of ricin would pose a serious threat to life. This is a conservative estimate, as the route of exposure and the person's immune status would also play a role.

The ricin molecule consists of two long chains, A and B, neither of which in itself constitutes a dangerous toxin—indeed other plants such as barley make chain A, without a noticeable effect on those who eat it. Chain B is the key to the toxicity. This binds to a particular carbohydrate component on the outside of a cell membrane and lies in wait. Eventually the membrane allows chain A to pass through, but once inside the cell it seeks out the site where the essential enzymes the cell needs are made and blocks it. Without its supply of enzymes the cell dies. A single ricin molecule is sufficient to kill a living cell, so in theory a mere $3\,\mu g$ of ricin, with its ten trillion molecules, would be enough to poison every cell of the human body.

It has been estimated that it would take about 10 castor beans to kill a human, and the same is true for other domestic animals. On the other hand, hens and ducks seem to be relatively unaffected when they eat this number of crushed seeds, in fact it takes ten times as many beans to kill them, and one bird, the tambourine dove, happily consumes the seeds of the castor plant without harm. Humans are particularly sensitive to the toxin, and it has been said that a single bean contains enough ricin to kill a young child, although this is an extremely rare event, and certainly in the UK it is almost unknown for a child to be poisoned by swallowing a castor bean. Even when this has happened, the child invariably survives the ordeal. What protects a child from the toxin is the seed coating of the castor bean, which is so hard that provided it remains unbroken it can pass through the intestine undigested and emerge intact.

The fatal dose of ricin for an adult is calculated to be as little as 70 µg, ten times the amount estimated from its LD_{50}. In practice a higher dose than this is needed to kill, because the immune system begins to produce antibodies which destroy the ricin invader, and it also depends on the route by which ricin has entered the body, *e.g.* whether by injection, digestion or inhalation. In fact between 200 and 500 µg of ricin is needed to kill an adult, but this is still a minute amount and several such doses would fit on the head of a pin. A larger dose is needed to kill if the ricin is swallowed rather than inhaled or injected.

1.3 Ricin in Chemical Warfare

There is no antidote for ricin, but there are vaccines available to people who are likely to encounter the toxin. These have been developed because ricin has potential as a chemical warfare agent, and indeed may have been used as such in the Iran–Iraq war of the 1980s. That the Iraqis possessed this capability was confirmed in 1995 when they admitted to UN inspectors that 10 litres of concentrated ricin solution had been produced. They also admitted that it had been tested as an offensive weapon in the form of artillery shells which would explode and disperse it. There was also evidence that they had continued to stockpile ricin, and in 1998 the RAF and USAF attacked a castor oil production plant in Fallujah, Iraq, which was suspected of producing it. That there was a continuing threat from ricin seemed to be confirmed when quantities of it were found in al-Qaeda caves in Afghanistan.

It had been known for many years that ricin could be a possible weapon and work on it by the US Army began in World War I. Eventually they concluded that it offered no advantages over the traditional chemical warfare agents, chlorine and mustard gas, which were then in use. Nevertheless it was developed further in World War II when cluster bomblets were tested as a way of dispersing a cloud of ricin dust over an enemy, but again the conclusion was reached that it was not very effective. The dust was meant to contaminate clothing and surroundings so that even after the attack it would be possible for a person to ingest a fatal dose merely by breathing it in. It would remain a threat for quite some time until the bond between chains A and B was severed by chemical reaction with components of the atmosphere, such as ozone and nitrogen dioxide. Once A and B become separate entities then the potency is lost. The difficulties with ricin as a weapon were, firstly, the need to protect one's own troops from exposure to it and, secondly, the inability to treat them if ever this happened.

Today, protection against ricin is possible but it has to be given before exposure to the toxin, in other words as a vaccine. There is still no antidote, nor is there ever likely to be. In 2003 the defence division of the US company, BioPharma, was licensed in the USA to test a genetically modified vaccine incorporating chain B, which induces the body's immune system to produce antibodies against the whole ricin molecule.

1.4   Production and Applications

Ricin is produced naturally by a number of plants but in such tiny amounts that it poses no threat, whereas the castor plant makes it in relative abundance. This plant is grown as a crop for its oil, which is extracted by heating the beans to 140 °C for around 20 minutes to denature the ricin proteins, and then crushing and pressing the beans to extract the 50% or so of oil which they contain. More than a million tonnes of castor oil are produced every year, mainly from beans grown in Brazil, India and China. Castor oil consists almost wholly of just one relatively rare fatty acid, **ricinoleic acid**. Indeed the presence of this fatty acid in a blend of edible oils can be used to detect whether any castor oil is present. Castor oil is used for preserving wood and leather, and as a raw material for the production of sebacic acid, which is used in brake fluids and as an engineering oil.

In ancient times castor oil was burned in lamps or applied as an ointment to the skin. It was used in these ways in India as long ago as 2000 BC, and its laxative properties were also recognised. A spoonful of castor oil was prescribed as a cure for stubborn constipation, which it did indeed cure. This was partly due to the traces of ricin which it contained, but at a cost to the patient in terms of stomach cramps and gripes—indeed it could lead to severe diarrhoea and dehydration.

For this reason it was used as a punishment in Nazi concentration camps, where prisoners would be forced to drink a cup of the liquid and suffer accordingly, sometimes dying as a result.

Ricin can be made from the waste left after castor beans have been cold pressed but it is a complex process involving a number of chemicals. It was first extracted from this source in 1888 by Hermann Stillmark and he named it ricin after the plant's botanical name, which itself was derived from the Latin word for a tick, *ricinus*, because of the supposed resemblance of the seeds to these insects. A process for producing ricin was even patented by the US Army in 1962, although the patent was removed from public view in 2004 because it showed those of evil intent just how easy it was to produce the toxin using simple chemical reagents such as sulfuric acid.

1.5 Ricin Poisoning

The effects of ricin poisoning depend on whether it is inhaled, ingested or injected, but by whichever route it enters the body many organs will be affected. Initial symptoms of ricin poisoning begin to appear after around six hours. After inhaling a deadly dose, breathing becomes difficult and is accompanied by high temperature, coughing, and tightness in the chest. As fluid builds up in the lungs breathing becomes more and more difficult and the skin takes on a bluish pallor as the blood becomes deoxygenated. Death will eventually ensue. If a toxic dose of ricin has been swallowed then the person will eventually vomit and suffer diarrhoea and there will be signs of internal bleeding. The patient will experience severe dehydration and reduced blood pressure, and within a day or two of exposure, the liver, spleen and kidneys will stop working. If the ricin has been injected, death will be even quicker and could take place within 36 to 72 hours of exposure, depending on the dose. If the dose is truly minute and the victim has been able to survive for five days, then recovery is likely.

Because no antidote exists for ricin, the most important factor is to avoid ricin exposure in the first place. If exposure to ricin dust has occurred, the affected person should remove all their clothing and shower as quickly as possible, and then seek medical help. If it is known that some ricin has been absorbed, then the victim needs supportive medical care to minimise its toxic effects. He or she will need help with breathing and be given intravenous fluids to counter dehydration, together with medication to treat symptoms such as seizure and low blood pressure. If the ricin has been ingested recently, then pumping and washing of the stomach should be carried out using a slurry of activated charcoal.

1.6   Detection and Identification

The symptoms of ricin poisoning are often mistaken for other diseases and proper diagnosis can thus be delayed for several days. By this time it may be too late to save the victim. If it is known that the patient has been exposed to the toxin, then diagnosis can be confirmed quickly. The presence of ricin in contaminated materials can be positively identified using enzymes, and simple testing kits are available to do this. The US Environmental Protection Agency (**EPA**) established the Environmental Technology Verification Program to help with their development and to verify that they worked. A typical device is the AbraTox kit produced by the Advanced Monitoring Systems Center, employing the bioluminescent bacterium *Vibrio fischer*, which as its name implies emits light. When it encounters toxins in a sample of water the light emitted is diminished in direct proportion to the amount of toxin present, and it can detect as little as 15 ppm of ricin. (It will similarly detect nicotine, Botulinus toxin, cyanide, nerve gases and V-agents.)

Another detection method relies on strips that change colour. A sample of suspected water is applied to one end of the strip and as it diffuses along the strip it passes several stripes which have been impregnated with specific antibodies, labelled with coloured dyes. If they change colour they not only show that a toxin is present but identify which one it is. In some test kits the antibodies are labelled with fluorescent markers which reveal a change when exposed to light. BioThreat Alert is one such product and it can detect ricin at 50 ppm or less. (It is able to detect *Botulinus* toxin as low as 10 ppb.) A range of such sensors and test kits is now available, some much more sophisticated than others and costing as much as $20 000, and most will reveal the presence of ricin within 15 minutes. On the other hand the cost of simple test strips is relatively low. For example, there are strips that cost as little as $250 for 10 and which can detect ricin at 20 ppm. Even more sophisticated technology using fibre-optic biosensors is now available and relies on a fluorescent-labelled anti-ricin antibody, IgG, immobilised on the surface of the fibre. As it encounters ricin molecules it transmits the variation in light intensity back to instruments for a more detailed assessment. It does this with no loss of intensity, which is in any case faint, and slight changes of which are normally difficult to measure accurately.

Thanks to all these developments the idea of using ricin to contaminate a water supply is no longer a serious threat, since a ricin alert can be triggered automatically. This was not the case back in the 1970s when the KGB investigated it as an assassination weapon. Detection in those days was still extremely difficult.

1.7 Positive Factors

Ricin is not all bad,—indeed it may one day help to heal rather then kill. Research has shown that it may have some beneficial medical uses, possibly to kill cancer cells. Chain A of ricin linked to antibodies has been researched as a possible way of achieving this. In one example antibodies have been designed so that they are attracted to cancer cells, where they destroy the blood vessels on which the cells rely and so cause the cancer to wither away. Such immunotoxins have been shown to work in the laboratory, but have yet to be tested on living things.

1.8 Examples of Ricin Attacks

Dissemination of ricin as a dust *via* the mail system is one way to use it to attack people. This has been tried, but has so far proved unsuccessful.

Someone in the USA, as yet unidentified, posted three letters containing powdered ricin, the first in October 2003, the second in November 2003, and the third in February 2004. In the first of these the ricin was in a sealed container inside a package sent to Greenville, South Carolina, the second was in a letter addressed to the White House, and the third in a letter to a leading senator's office in Washington DC. It was the last of these which caused the most alarm—the FBI kept knowledge of the first two out of the media. The third letter, addressed to Republican Senator Bill Frist, arrived on 2 February 2004 when he was Senate Majority Leader. As it passed through a mail-opening machine some white powder fell out but for some time went unnoticed. When it was observed, its discovery immediately suggested a repetition of the anthrax scare of 2001 when spores of this pathogen were disseminated as a powder *via* the US mail.

This ricin story began on 15 October 2003 when a package addressed to the Department of Transportation was intercepted at the sorting office which served the Greenville–Spartanburg airport, South Carolina. The package was labelled with a warning which read:

> **Caution RICIN POISON enclosed . . . do not open without proper protection**

Inside was a small vial and a note signed 'Fallen Angel' which appeared to be from the owner of a fleet of tankers. Whoever it was, they demanded that new federal regulations reducing the hours which long-distance truck drivers could work should be withdrawn. If this did not happen by January 2004, when they were due to come into force, then Fallen Angel threatened to send more ricin-laden letters through the post. Inside the package was a sealed container and in it was a white powder which tested positive for ricin, but it was of such poor quality it really posed a threat to no one. Federal agents tried to discover the sender by accessing the drivers' logs and time sheets of nine trucking companies who made deliveries in the Greenville area but nothing came of their investigations.

The next ricin-laced letter was addressed to the White House and had been opened at an offsite mail processing facility, where all such mail is subject to irradiation in order to destroy any possible pathogens. This safety procedure was instituted after the anthrax attack on various media organisations in the autumn of 2001, when letters containing anthrax spores were sent to Boca Raton in Florida, to New York City, and to the offices of two senators in Washington. These attacks killed five people and the nationwide alert disrupted the US postal service for several weeks. The person responsible was never found, or at least was never brought to court.

The second ricin letter arrived on 6 November and also contained impure ricin, with a note which threatened more attacks:

> **I will turn D.C. into a ghost town. The powder in the letter is RICIN. Have a nice day, Fallen Angel.**

This letter was not made public until after the third ricin letter was received three months later. This third letter resulted in three

Senate office buildings being closed down for a week, of which the one most affected was the Dirksen Senate Office Building, where the substance was found on the afternoon of Monday 2 February by a young worker in Senator Frist's fourth-floor mailroom. Suspicion that this was another anthrax attack meant that 16 staff from that floor were immediately quarantined and decontamination was undertaken. Health officials urged other staff to report immediately if they developed a fever or cough in the next two or three days, but in any case 30 of them went through the decontamination process and the buildings were sealed off.

Specially trained personnel in full protective gear inspected the offices but found nothing, and the offices were reopened the following week. The FBI and other agencies conducted tests and quickly identified the powder as impure ricin. This was done at Fort Detrick, Maryland, where army scientists used an electron microscope to determine the size of the ricin particles, which is crucial to determining whether any ricin might have been inhaled. It proved impossible to trace the sender of the letter because it was uncertain which piece of mail the ricin had fallen from, and this time no threatening letter was found. During February the FBI announced a $100 000 reward for information that would catch the person responsible. As far as is known this has never been claimed.

The US Government already knew that ricin might be used against it. A few years previously they had arrested Ahmed Ressam, who had been trained as an al-Qaeda terrorist. He was convicted of plotting to blow up Los Angeles airport on New Year's Eve in 1999, and he said at his trial that the militant Islamist group to which he belonged were planning to use ricin smeared on door handles as a weapon. Whether they ever did this is unknown.

Across the Atlantic, the UK Secret Service thought they had uncovered a plot to use ricin. On 5 January 2003 a group of six supposed terrorists of North African origin were arrested in a police raid on a flat over a pharmacy in the North London suburb of Wood Green. The media were told that they were part of an al-Qaeda group, and were in possession of ricin and were in course of preparing further quantities of it. Alarming reports appeared in newspapers of how this toxin could be spread as a powder on the London Underground or added to reservoirs, of which there are several large examples close to Wood Green. The news flashed around the world and was even referred to by Colin Powell, the US

Secretary of State, when making the case for military intervention in Iraq, from where it was claimed that some of the ricin had originated. The UK authorities had been tipped off by the police in Algeria, who had obtained the information while questioning a suspected terrorist, Mohamed Meguerba.

Meguerba had been trained at an al-Qaeda camp in Afghanistan and tried to enter Europe on a false passport, which led to his arrest at Amsterdam airport. He was later released and made his way to the UK where he was again arrested, this time for credit card offences, but he was once more released on bail after appearing to be unwell. He fled the country, eventually ending up in Algeria where he was detained and where he revealed details of the Wood Green ricin plot. His fingerprints were found at the gang's Wood Green hideout.

Items from the flat were taken away and sent for analysis to the UK Government Chemical Defence Establishment at Porton Down. These items included 20 castor beans, a coffee grinder, a funnel, weighing scales, rubber gloves, and instructions on how to extract ricin. Somewhat oddly, a large number of apple pips were also found. Tests at Porton Down showed that the would-be terrorists had been completely unsuccessful in isolating any toxin.

Meanwhile other suspected militant Islamists were sought and a week later five men were arrested in Bournemouth and three in Manchester. One of them was Kamel Bourgass,[i] who had evaded capture at Wood Green and had fled northwards to the gang's safe house in Manchester, where he was cornered but was not prepared to give up without a fight. It was during this that he knifed and fatally wounded police constable Stephen Oake, for which he was eventually brought to trial. On 30 June 2004 he was found guilty of the murder and sentenced to life imprisonment, to serve a minimum of 20 years. The other members of the Wood Green terrorist cell who were brought to trial at the same time had to be released because the evidence against them was insufficient to secure a conviction. All that could be said was that maybe they might have planned to extract ricin, but they denied that this was their intention. Nevertheless, they were found guilty of possessing false immigration papers and were deported.

---

[i] Real name, Nadir Habra.

However, back in January 2003, the terror alert appeared justified when samples of a powder found in the house were analysed at Porton Down and confirmed to be ricin. This was a mistake, however, as it later transpired that nothing found at the Wood Green flat had in fact been contaminated with ricin. It appears that the original information came from a preliminary report of the analyst's findings which should have read that the presence of ricin had *not* been confirmed. The 'not' had somehow been omitted, or so it was later claimed.

A year earlier in the US, ricin had been confirmed in an attempted murder case. Kenneth Olsen worked for Agilent Technologies, a company specialising in sophisticated measurement techniques, including bioanalysis, and based at Liberty Lake, Spokane County, Washington. Olsen secretly planned to produce ricin with a view to disposing of his wife. To this end he downloaded information from the internet which explained how to kill with undetectable poisons and in particular how to extract and dispense ricin. He bought some castor seeds and in his workspace he began to process them. By the time of his arrest on 19 June 2002 he had managed to obtain three grams of a white powder from the seeds. When this was sent to the US Army Medical Research Institute for analysis it was shown to contain 20% ricin. Olsen had made 600 mg of ricin, enough to kill at least 100 people. He was arrested on a federal charge of possessing a biological weapon with intent to use it. A year later, in July 2003, he was convicted of this charge and sentenced to 13 years in prison and fined $23 000. The fine was calculated as the cost of cleaning up what was deemed to be hazardous waste contamination in his workplace, car and home.

As a molecule for murdering someone, ricin was not living up to the reputation it had gained many years before, when it had been successfully used in a famous case of political assassination.

1.9 The Murder of Georgi Markov

Georgi Markov died in intensive care in a London hospital on Monday, 12 September 1978. The doctors treating him believed his death was due to natural causes, in particular to a virulent form of blood poisoning. Others suspected that he had died an unnatural death, however, and speculated about this in the media. They were proved right when the pathologist who carried out the post-mortem examination probed the curious hole in his leg and retrieved from it a tiny pellet. It was soon clear that his death was not due to natural causes—in fact he had been assassinated.

By a strange twist of fate the assassination attempt probably occurred within sight of my office at King's College, which overlooked the River Thames and Waterloo Bridge. I lectured in inorganic chemistry in the Chemistry Department, which occupied the upper floors and had spectacular views over the river Thames on one side and overlooked the Strand and Bush House on the other.[ii] One morning in late September 1978 a reporter from *The Guardian* newspaper, David Pallister, rang me and asked if I could suggest which poison might have killed Markov. Only a month previously I had written an article for the popular science magazine *New Scientist* about thallium, the toxic metal which a serial poisoner, Graham Young, had used to kill his workmates and which had been in the news. I was more concerned to tell of the history of thallium and its new use as a reagent in chemical synthesis, but it was my apparent knowledge of its toxicity which brought the article to David's attention. I protested that I could be of little help, but he had by then obtained an enlarged photograph of the pellet that had killed Markov and thought that it might be possible to deduce what the poison was from its dimensions. Intrigued, I invited him round to King's College.

We sat down together in my office and began a simple form of analysis *à la* Sherlock Holmes. Our first task was to calculate the volume of the two holes that had been drilled as receptacles to hold the poison. The pellet had a diameter of 1.7 mm and the two holes each had a diameter of 0.4 mm; the total volume of the two cavities was 0.4 microlitres, and we estimated that these could hold at most 0.4 mg of the poison. It was now a question of what substances could be fatal to an average adult at such a low dose. This tiny amount immediately ruled out inorganic poisons like arsenic and thallium, where the dose has to be 100 mg or more. It also ruled out organic poisons such as nicotine or fluoroacetate which, while more than twice as deadly, would still need several mg to cause death. It even ruled out nerve gases such as sarin, of which 1 mg can kill. Radioactive poisons were a possible contender, as we shall see in Chapter 10, but the speed of Markov's demise ruled out that possibility.

[ii] The Chemistry Department closed its doors in 2005.

All that now remained were biotoxins of the kind produced by snakes, such as the α-bungarotoxin produced by cobras, or those produced by bacteria, such as the deadly *Botulinus* toxin. I consulted a fellow academic, a toxicologist in the Department of Anatomy,[iii] and he confirmed that we were on the right track and named the above toxins as the most likely. Our speculations appeared in the *New Scientist* issue of 12 October 1978, and we were correct in that it was a lethal biotoxin, but wrong in our choice of which one it might be. The poison was identified later that month as ricin, which is twice as deadly as cobra venom but less deadly than *Botulinus* toxin.

Both the poison and the weapon were a unique combination for an assassin and it is this factor which has kept interest in Markov's murder alive. Back in the 1950s and 60s the traditional city gent in London was dressed in a pinstripe suit, wore a bowler hat, carried a black briefcase, and walked with a regulation black rolled umbrella. Such a sight began to disappear in the 1970s and is now hardly ever seen, but back in 1978 it was still not uncommon to see men with umbrellas walking to work or standing at bus stops. In 1978 the Bulgarian Secret Service decided that they could disguise an air rifle or gas gun as an umbrella or a brief case, and that this could fire a pellet with enough velocity to penetrate clothing and bury itself beneath the skin, there to deliver a fatal dose of a biotoxin. The pellet might even go undetected. In August 1978 their plan was put into operation and the targets were Vladimir Kostov, who lived in Paris, and Georgi Markov, who lived in London. Kostov survived the assassination attack, but Markov was murdered. No one was ever brought to justice for the crime, but those who instigated it, organised it and carried it out have almost certainly been identified, as we shall see.

The USSR Secret Service, then known as the KGB, is thought to have made the ricin that killed Markov, indeed there is some evidence that this organisation also made use of ricin to harass their own dissidents who had fled to the West, such as the highly influential anti-Soviet author Aleksandr Solzhenitsyn. He believed he had been poisoned with it in 1971. Markov certainly was—and in his case fatally.

[iii] Who wishes to remain anonymous.

Georgi Markov was born on 1 March 1929 in Sofia, the capital of Bulgaria, and died on 11 September 1978 in England. He was buried in the graveyard of St Candida and St Cross Church in the quiet village of Whitchurch Canonicorum at the request of his English wife, Annabel, whose family came from that part of Dorset. The church is unique in the UK in that it still has an intact shrine which holds the bones of St Candida, a Christian martyr who died in a massacre in the Netherlands on 5 June 755. The vicar quoted one of Markov's poems at the burial service:

My funeral was as merry as can be;
The birds sang and chattered from a tree.
My friends in joy and mirth were clad;
Only I felt a little sad.

Markov was a novelist and playwright and this kind of irony was his trademark. In his communist homeland he had become famous when his first full length novel, *Men*, was published in 1962 to great acclaim, winning the annual award of the Union of Bulgarian Writers, and it brought him fame throughout the Eastern Bloc of communist states allied to Soviet Russia. A succession of successful novels, of which the best known were *A Portrait of My Double* and *The Women of Warsaw*, made him rich, rich enough to own a BMW, a car normally reserved for leading members of the ruling party. He also wrote plays, although few were ever performed because they were not approved by the all-powerful Cultural Committee, yet Markov was on friendly terms with country's president, Todor Zhivkov.

Despite the advantages he enjoyed in his homeland, Markov desired the kind of freedom that writers enjoyed in the West, and he hoped that this would become possible because of what was happening in the late 1960s when the repressive communist regime in Czechoslovakia gave way to a new spirit of openness. Consequently he became very unhappy when the freedom movement was violently suppressed by the Soviet military in 1968. Markov's next play expressed his disapproval and he was summoned before the Government's Cultural Committee to explain his openly revisionist attitude, and to be told that his new play would not be produced. Nevertheless, he continued to have some political influence and in 1969 he obtained a rarely-bestowed visa allowing him to visit his brother in Italy. However, when he applied for an extension the Bulgarian authorities refused, so he decided to remain in the West. In Bulgaria he was tried *in absentia* for this crime and sentenced to six years in prison, and his books were ordered to be removed from the shelves of public libraries. Markov now moved to London, learned English and even translated some of his plays, one of which, *The Archangel Michael*, won an award at the Edinburgh International Festival. He joined the Bulgarian section of the BBC World Service in 1971. There he was employed because of his language skills, but he was not allowed to express his personal opinions over the airwaves.

Markov also regularly went to Germany to broadcast, in particular to Munich to the CIA-funded Radio Free Europe. There he was encouraged to use his knowledge of Bulgaria and its ruling elite to ridicule and to taunt them for their subservience to the USSR. Normally such behaviour would have ended his employment with the BBC, whose output has always to be non-political, but the Bulgarian section of the World Service pleaded for him to be retained because they could ill afford to lose his talents. On Radio Free Europe, Markov was allowed a 20-minute slot every Sunday evening in which he talked about real life in Bulgaria, based on his personal knowledge of the country and its corrupt regime.

Vladimir Kostov also worked for Radio Free Europe and he too had been well connected to the communist leadership in Bulgaria before he defected in 1976. Kostov has been the former head of the Paris Bureau of the Bulgarian State Radio and TV network and had applied for, and been granted, political asylum by the French Government in June 1977.

Although the Bulgarian government jammed all Radio Free Europe broadcasts it seems highly likely that they recorded them in one of their Western European embassies, and it was popularly joked that they were avidly listened to by the communist authorities. Not surprisingly, they did not like what they heard and it appears that someone in the Bulgarian security service demanded that Markov and Kostov had to be silenced.

The Bulgarian Secret Service's trained assassin struck at Kostov in Paris on 27 August 1978. He was attacked as he was with his wife on the escalator at the Arc de Triomphe station of the Paris Metro. A man manoeuvred himself to be standing immediately behind Kostov as he ascended to the station exit. As they neared the top of the escalator the man fired a pellet into the small of Kostov's back. Kostov said afterwards that he had noticed a man who appeared to be shadowing him when the couple were on the underground train. On the escalator Kostov felt a prick in the small of his back just above the belt of his trousers, and he heard a muffled sound like that of an air gun. He turned and saw it was the man he had seen earlier, and that he was carrying a small briefcase. The man moved quickly away and was lost in the crowd of people leaving the underground station. The day after he had been shot Kostov developed a fever and noticed swelling around the point where the pellet had pierced his skin. He went to see a doctor, who diagnosed a wasp sting, but when his condition worsened he went to a local hospital. The doctor there said that he was in no grave danger, and indeed his temperature began to go down and within a few days he was almost back to normal. Although he did not know it, he was now walking around with a dangerous dose of ricin in his body.

After Markov's death Kostov allowed doctors to remove a sample of flesh from his back at the point where the pellet had lodged, and this was sent to the UK both for forensic examination and for testing at the research centre at Porton Down. Not only did it have the same tell-tale holes, the pellet also held some ricin and antibodies against ricin were detected in the flesh. It was clear that only a small fraction of the ricin had escaped from the pellet and that Kostov's immune system had had time to develop some protection against the toxin, thereby helping him to survive. What also saved him was the layer of wax coating the pellet, which had not all dissolved in his tissue. This meant that that most of the estimated 0.45 mg of ricin the pellet had contained remained in place.

The Markov story turned out to be more complex than Kostov's in that he had been warned of his impending assassination by his brother Nikola, a stamp dealer in Bologna, Italy. He later said that in the early summer he had been visited by a person he did not know, but who had told him what the Bulgarian secret police planned to do to his brother Georgi. He passed on the information, but what could Georgi do? He simply carried on as normal.

On the fateful Wednesday, 7 September 1978, Markov had driven his car to work as normal and parked it in an all-day parking lot near the National Theatre, on the South Bank area of the Thames. He then went to the southern end of Waterloo Bridge to catch a bus to take him across the river to Bush House about half a mile away. As he waited, he suddenly felt a sharp pain in his right leg and turned round to see what had caused it. It appeared to have been a prod from the umbrella of a man who was now retrieving it from the pavement. The man mumbled an apology, and then hailed a passing taxi. He was about 40 and swarthy looking and spoke with such a broken English accent that the taxi driver had difficulty in understanding him. Markov told a colleague at the BBC, Teo Lirkoff, about the incident with the umbrella and showed him a spot of blood on the back of his jeans.

It is possible that the weapon the assassin used to kill Markov was similar to that disguised as a briefcase with which he attacked Kostov. Indeed, after he had been attacked Markov said that he thought his attacker had dropped his umbrella merely as a way of hiding his face. It would have looked rather odd for the assassin to shoot Kostov in the back with an umbrella while riding the escalator, but a briefcase would have been a much less obtrusive object. It would seem likely that the same man attacked both victims, but why would he use two different weapons? One possibility is that he might have realised that the air gun in the briefcase had not been powerful enough, because Kostov had not died, and he therefore resorted to the umbrella gun which he could direct at a different part of the body, in this case the back of a thigh.

Later that day Markov returned to his home in Lynette Avenue, south of Clapham Common, arriving at around 10:30 pm. That night he slept in his study because the following day he had to leave early and did not want to wake his wife, Annabel.[iv] However, at about 2 am she heard him vomiting and found he was feverish. When she took his temperature it was 104 °F, so she telephoned her doctor who said that Georgi probably had the flu and to keep him cool. Annabel sat with her husband all that night, during which he described to her the incident in the bus queue and said that he wondered if that had anything to do with his worsening condition.

The next morning Annabel went off to work at 10:30 am and took their daughter with her so that Georgi could stay in bed and get some much needed sleep. Later that day the doctor called to see him and found his condition had deteriorated so badly that he rang for an ambulance and had Georgi taken to the nearby St James's Hospital in Balham. The doctor who examined him there was

---

[iv] They were married in 1976 and had a two-year-old baby daughter.

Bernard Riley, the house physician, and when Georgi told him about the mark on his leg the doctor examined it and noted that it had become inflamed, with a central puncture mark, but he assured Georgi that it was most unlikely to be the cause of his condition. In any case the puncture mark on his leg was only about 2 mm wide and far too small to be the result of a prod from the tip of an umbrella. Markov insisted to Riley that he had been shot in the leg with a lethal dart, but an X-ray revealed nothing unusual. In fact the pellet was later found on the X-ray film, but had been so small that the radiologist assumed it was nothing more than a blemish on the film.

All that day Markov continued to deteriorate. His blood pressure fell rapidly, while his white blood cell count rose to 33 000, the highest that Ripley had ever witnessed in a patient. (The normal range is 5000 to 10 000.) The doctor diagnosed that Markov was suffering from severe septicaemia. On the Saturday Markov's blood pressure fell dramatically and his pulse was racing, registering 160. He now became cold and dizzy and was transferred to intensive care. He was still vomiting and blood was present. He had also stopped passing urine, a sure sign that his kidneys were badly damaged. On the Sunday an ECG showed damage to his heart and as the day progressed he became worse. Early on Monday morning he became delirious and began pulling out his intravenous therapy tubes. His heart stopped at around 10 am and attempts to revive him were without success; he was declared dead at 10:40 am. His post-mortem showed damage to a number of organs, particularly his lungs, liver, intestines, lymph glands, pancreas and testicles, and there were haemorrhages in several of his tissues. It was concluded that he had in fact died of what the doctors had suspected, namely acute blood poisoning.

In his last broadcast on Radio Free Europe, which went out on Sunday, 11 September, Markov poked fun at the dates of two back-to-back public holidays in his native land: one was the Day of National Freedom which was celebrated on 9 September, and the other was the Day of the Militia and State Security, whose ceremonies were held on 10 September. Markov's assassin had struck earlier that week, so no doubt the Bulgarian authorities knew that this time the joke was on him. In fact he died the following day.

Georgi Markov's inquest was held at Battersea on Tuesday, 2 January 1979 by Gavin Thurston, the Inner West London Coroner.

It lasted two hours and arrived at the verdict that Markov had "been unlawfully killed by ricin poisoning when a metallic pellet was inserted into the back of his thigh." The coroner based his verdict on the evidence supplied by expert witnesses.

Dr Rufus Crompton was the pathologist who carried out the post-mortem examination on Markov, who had been 6 ft tall and weighed 14 stones,[v] and he deduced that Markov had died of acute toxaemia. Although he could not deduce what poison or toxin had caused it, it was linked to the metallic sphere which he extracted from the back of Markov's right thigh, and which he said was located just below the skin. The pellet and a sample of thigh tissue surrounding it were sent for testing at the Chemical Defence Establishment at Porton Down in Wiltshire, and the results were reported by Dr David Gall. He said the toxin was most likely to have been ricin. When a pig was injected with a dose of the size that would have fitted within the pellet the animal suffered in the same way that Markov had, and with almost identical symptoms, namely fever, a very high white blood cell count and collapse of the cardiovascular system. It had died even sooner, after 24 h. A post-mortem on the animal revealed that it too had suffered multiple haemorrhages to the internal organs, just as Markov had.

Although Gall could not detect ricin antibodies in Markov's remains, they were detectable in the sample provided by Kostov which had been sent to the Metropolitan Police Forensic Laboratory, and thence to Porton Down, where pathologist Gall and his colleague Dennis Swanson were asked to investigate. Gall later admitted that they had almost lost the platinum ball, which had worked its way to the surface and was brushed away when he ran his hand over the flesh!

The pellets were returned to the Police Forensic Science Laboratory, where electron microscopy determined their exact dimensions. Other tests showed them to be an alloy of platinum (90%) and iridium (10%), which is one the human body is known not to reject. This alloy is also inert to ricin. At first it was thought the pellets were of the type used in the tips of ball point pens but the manufacturers confirmed that this was not the case.

Forensic inspection of Markov's jeans revealed no trace of a firearm powder, nor evidence that they had been pierced by a

---

[v] Equivalent to 196 lb, or 89 kg.

hypodermic needle, which was one possible way to inject the pellet. It was consequently deduced that it had been fired from a gas or air gun.

So who killed Markov? The media had given Markov's death a great deal of publicity, the novelty of a poisoned umbrella being the feature of interest, and naturally they pointed the finger of suspicion at the Bulgarian Secret Service. The Bulgarian Government countered these claims, saying that he had been eliminated by the CIA who had no further use for him at Radio Free Europe.

The person responsible for ordering the assassination of Markov was the Bulgarian Interior Minister, Dimiter Stoyanov, and the plan was put in operation by his deputy, General Stoyan Savov, helped by the KGB. Apparently there were three attempts on Markov's life, the first two being designed to poison his food and drink, but they had failed.

In 1990 a double agent, Oleg Gordievsky, who had worked for both the KGB and MI5, revealed in his memoirs that it was the Soviets who provided and coached the Bulgarian secret police how to kill using a ricin-charged pellet. Eventually, one person came into the frame as being the most likely assassin and he was the Dane, Francesco Gullino, who was of Italian origin. He posed as an antiques dealer and it appears that he was caught smuggling currency by the Bulgarian customs and handed over to the secret police, who realised that he could be important to them. He agreed to cooperate.

After the fall of the USSR lots of information came to light, including the identity of the secret laboratory where the umbrella was made. The collapse of communism in Bulgaria also revealed much about the assassinations, including the code name that agent Gullino used, which was Piccadilly. In February 1993 he was arrested in Copenhagen, questioned and fingerprinted and he admitted he had been a secret agent, but not an assassin. The previous year the former head of Bulgarian intelligence had been arrested and sentenced to 16 months in jail for destroying an archive of 10 volumes which gave details of assassinations ordered by the regime; this has ensured that it will never be known for certain who killed Markov. Savov, the deputy interior minister at the time of Markov's death, had committed suicide the previous year while facing trial for his crimes. Justice of a sort had been done.

CHAPTER 2

Hyoscine and the Murder of Belle Elmore (Mrs Crippen)

A word shown in **bold** *indicates that further information will be found in the Glossary*

2.1 The Unlucky Mrs Crippen

At 9 am on 23 November 1910 the public executioner, John Ellis, whose normal mode of employment was as a men's hairdresser, dispatched Dr Hawley Harvey Crippen to the next world. Crippen had been found guilty of murdering his wife, Belle Elmore, by poisoning her. The crime fascinated the public for several reasons, and it has retained this fascination ever since. First, Belle was a minor celebrity, a popular singer of her day. Second, her body was dissected and the flesh buried under the cellar floor; but her bones and head were never found. Third, she was murdered because her husband had a mistress, Ethel le Neve, whom he wished to marry—divorce was unthinkable. Fourth, Crippen and Ethel tried to escape across the Atlantic by posing as father and son. Fifth, their deception was uncovered by the captain who telegraphed his suspicions back to Britain with the result that they were arrested as they disembarked in Canada. Finally, and most importantly from our point of view, Belle was murdered with an unusual poison: **hyoscine**. Crippen must have had a reason for choosing this particular agent, and it gives us a clue as to how he hoped to commit the crime and also why he suddenly had to resort to a rather messy Plan B.

Molecules of Murder: Criminal Molecules and Classic Cases
By John Emsley
© John Emsley, 2008

2.2   Hyoscine (Scopolamine)

This natural chemical was traditionally a secret drug taken by those who dabbled in the occult. Anyone who wished to make contact with the spirit world, the future, or their inner self, could call upon its powers in the form of potions made from certain leaves, roots or seeds. All over the world there are plants rich in hyoscine and folklore warned that they were dangerous to eat, but it was also believed that they could confer special powers on those who knew how to use them. For example, in Europe witches were reputed to use mandrake (*Mandragora officinarum*) in the form of a fatty extract which they were believed to daub under their arms and insert into their genital area. We now know that this would have been an effective way for the body to absorb the drug, and some modern hyoscine medications are in fact applied as patches to the skin. Having smeared on the extract, they might have imagined they could cast spells, or even that they could fly! That such nonsense was believed seems strange in the modern world, indeed the persecution and burning of supposed witches were spectacles of extreme cruelty. In other parts of the world, those who fell under the influence of hyoscine were treated more reverentially, as if they were in touch with angels rather than devils.

2.3   Medical Uses

So what does this natural chemical do? Its most obvious effects are on the brain and the central nervous system. Small doses lead to feelings of elation and a lack of inhibition. Larger doses lead to mental confusion, however, and the person may have hallucinations and talk wildly, a state which others in bygone ages believed to be providing prophetic visions and divinely inspired messages. Eventually hyoscine produces a deep and dreamless sleep from which the person will awake with no memory of what has happened. Too large a dose, of course, and the sleep lasts forever.

Hyoscine has the medical name scopolamine, derived from the plant genus *Scopolia*, the members of which contain it, and which is named after the Italian botanist Giovanni Antonio Scopoli (1723–1788). Hyoscine is an alkaloid and has a complex chemical name.[i] In its pure form it is a viscous liquid which is soluble in water, from which it can be crystallised as the hydrobromide. In this form hyoscine still has a place in modern medical treatment, as also has a chemically related form, the butyl bromide, which is sometimes preferred because it cannot penetrate the brain but has a localised calming effect on the muscles of the stomach and intestines.

Plants which produce hyoscine have long been known to be dangerous, and include the aptly named deadly nightshade (*Solanum dulcamara*), henbane (*Hyoscyamus niger*), jimsonweed (*Datura stramonium*), the various angel's trumpets (*Brugmansia*) of South America—and mandrake. Mandrake was the most notorious of these, as it was supposedly endowed with demon-like properties, maybe due to its curiously shaped root which looked almost as though it were part of a tiny human creature. The plant was supposed to give off a deadly shriek when pulled out of the ground, the sound of which could kill, or at least send a person mad. This strange belief is revived in J K Rowlings' *Harry Potter and the Chamber of Secrets,* but it is one of long standing, and is referred to by Juliet in Shakespeare's play *Romeo and Juliet.* She muses on what it will be like when, having been drugged so that she appears dead, she then recovers consciousness in the crypt where her body will have been laid to rest. She imagines the horror of it: ". . . loathsome smells and shrieks like mandrakes torn out of the earth, that living mortals, hearing them, run mad . . ."

Mandrake produces a yellow plum-like fruit which has a sweet taste and can be eaten without harm when it is fully ripe. It was thought to have aphrodisiac qualities, and the ancient Greek philosopher Theophrastus wrote of it as such in 300 BC. Apart from the ripe fruit, all parts of the plant are poisonous if eaten. In some societies mandrake was thought to ward off evil spirits, but in others it was regarded as evil *per se.* In 1630 in Hamburg, Germany, the very fact of possessing it was taken as proof positive that the person was a witch, and on this basis three women were executed for witchcraft. In earlier times mandrake root was believed to cure infertility. In Chapter 30 of *Genesis* is the story of Rachel, who

---

[i] (1S,3S,5R,6R,7S,8R)-6,7-epoxy-3-[(S)-tropoyloxy]tropane, which gives information about the chiral centres within the molecule – see Glossary.

desperately wanted a baby by her husband Jacob. He had already proved his manhood by siring children by Rachel's maid, and even by the maid of his sister-in-law Leah. He also had a son and a daughter by Leah herself. Rachel had agreed to that immoral liaison in order to obtain some of the mandrake that Leah's son had found in a field. This appears to have worked, because eventually Rachel was rewarded with a son of her own. Or so the story goes.

How Rachel used the plant is not revealed, and clearly she took only a little because otherwise it could have been dangerous. The same is true of any plant that produces hyoscine, as the early American colonists were to discover when they encountered the plant known as Jamestown weed in the area around Jamestown, Virginia. This is also known as stinkweed, devil's apple or thorn apple. It grows up to two metres tall and has white or purplish trumpet-like flowers with an off-putting smell. It produces a large fruit with spines, hence the name thorn apple; these contain around 0.75% hyoscine. In 1676 there was a rebellion in Jamestown led by Nathaniel Bacon and soon much of the colony was under his control. Troops were sent to restore order. Times were hard and when the rebels ran short of rations they ate the unfamiliar fruit, resulting in mass poisoning. Thankfully the troops were not needed, because the rebellion collapsed when Bacon suddenly died. Jimsonweed, as it is now called, flourishes best in a warm climate and it has even been grown as an ornamental plant in gardens in the south of England.

Henbane is another plant that produces significant amounts of hyoscine. Henbane was to be found across the Middle East from Egypt to India and it is grown in California as a crop which yields both hyoscine and **hyoscyamine**. Hyoscine exhibits **chirality**, in other words it can exist in two molecular forms which are mirror images of each other. These have very different effects within the human body, an observation that was first deduced by Arthur Cushny (1866–1926), a professor at the University of Michigan from 1893 to 1905. He studied the effect of the two optical isomers of hyoscine, first on animals and then on healthy volunteers, including Cushny himself and his assistant Alvin Peebles, and then on patients at the Michigan Asylum for the Insane at Kalamazoo. He compared the effectiveness of the two isomers in sleeping draughts by simply noting how long a recipient slept after taking a measured dose. The test was one of the first to be carried out along modern lines with the two forms being given randomly to patients, and including nights when they were given neither drug. Cushny found that one isomer, the L-isomer (L standing for the Latin *laevo*, meaning 'left'), was much better than the D-isomer (D being short for *dextro*, 'right') and he concluded rightly that this showed they targeted a specific receptor in the brain, the L-isomer being better fitted to the task.

Henbane seeds have the highest concentration of hyoscine and indeed the ancient Egyptians knew that they were deadly poisons. In Shakespeare's play *Hamlet*, the ghost of Hamlet's father tells his son how he was murdered by his uncle who poured a potion made from henbane into his ear while he was having his afternoon nap. "Sleeping within my orchard . . . thy uncle stole, with juice of cursed hebenon [henbane] in a vial, and in the porches of my ears did pour the leprous distilment . . ." is how the ghost tells it. Shakespeare even speculated on the reason why it was swift to act. It affected the blood so that it "doth posset and curd, like eager [acid] dropping into milk". In that respect he was wrong, but maybe not about this method of applying the poison, which seems strange because unless the eardrum was perforated the liquid would not enter the body. Nevertheless in 2002, Basilio Kotsias, a Buenos Aires doctor, expressed the opinion in the *Archives of Otolaryngology* that it would in fact have been possible for King Hamlet of Denmark to have been poisoned this way. He indicated that scopolamine could be absorbed into the body through the ear by diffusing through the tissue, and could deliver several mg of the drug into the blood and reach a toxic level.

The hyoscine molecule is dangerous because it antagonises many of the effects of the body's essential messenger molecule acetylcholine (ACh) which activates both the voluntary and involuntary muscles, the glands which secrete bodily fluids, and many brain functions. The brain must keep its supply of ACh high if it is to work properly. However, if the level of ACh is too high—and this can happen if the mechanisms for removing it are malfunctioning—this results in an unnecessary and distressing level of brain and bodily activity. (If too little ACh is produced, say in old age, then this equally can cause a problem.) There is therefore a need for a drug to counter the effects of ACh and regulate its action, and hysocine is just such an "anticholinergic" agent. Hyoscine distributes itself around the body with a half-life of between 5 and 10 hours, which means that a non-lethal dose does not pose a long-term threat. Research shows that after three days only about 3% of an injected dose remains. Any such injection has to be small, however, because as little as 15 mg delivered in this way can quickly kill.

ACh is a key part of the central nervous system and is produced and stored in small vessels, known as vesicles, at the ends of nerve fibres. When a signal arrives along the nerve ACh is released and crosses the synapse gap to the organ to be activated. There it acts rather like a key and locks into a receptor, switching it on. The receptor can be fooled into accepting a different molecule such as hyoscine, assuming it to be ACh, but while the new key may fit the lock it will not activate the mechanism. It blocks the receptor until it is eventually disgorged.

Doctors used to prescribe hyoscine hydrobromide to counteract dizziness, muscular spasms, and to correct an irregular heartbeat. Depending on the dose it can calm the brain or induce a deep sleep, and it especially affects the brain's ability to remember things. Research on hyoscine hydrobromide showed that an injection of 0.5 mg prevented short-term memory formation, becoming progressively worse until after an hour no new information was retained. The loss of memory is similar to that experienced by those suffering from Alzheimer's disease. Those who have died with this condition are found to have few, if any, ACh receptors active in the cerebral cortex of the brain, which is the region responsible for higher intellectual functions. This is exactly where hyoscine does its work.

Hyoscine slows down the movement of food through the stomach and intestines by calming the muscles, which is why it is prescribed for irritable bowel syndrome. There is evidence that doctors in ancient Egypt knew of its benefits in calming the gut as long ago as 1500 BC, as revealed in the Ebers Papyrus, recently researched by Jackie Campbell at the KNH Centre for Biomedical Egyptology, University of Manchester, UK. This document contains more than 800 remedies for all manner of ailments. Plants known to contain hyoscine were used in the treatment of diarrhoea, and would have slowed passage through the gut by influencing the muscle action of the intestines.

Not unexpectedly, one of the side effects of taking the drug can be constipation. Other side effects of hyoscine are a dry mouth and itchy eyes, because it immobilises the membranes releasing fluids which assist digestion and lubrication. For this reason it has been used prior to surgery to dry up bodily secretions, such as those in the lungs, nose and mouth, in which case it would be given as an injection at around 0.5 mg.

Larger doses of hyoscine can cause stupor, paralysis and delusions, and even when used medically its side effects can be serious, producing rapid breathing, excitability, dizziness and hallucinations. Hyoscine is a psychedelic deliriant drug, and it may even be possible to become addicted to it. A case was reported in 1960 in which a 54-year-old woman became hooked on hyoscine-containing sea-sickness tablets, consuming a dozen or more every day. As a result she became neurotically depressed and prone to hysteria.

Despite its somewhat unpredictable effects, hyoscine continues to be used medically—and indeed criminally. Medically, hyoscine calms the central nervous system, which is why it can act as a sedative and is able to control bodily spasms of the kind associated with manic behaviour and Parkinson's disease. Because it so obviously has an effect on the brain it was at one time injected along with morphine as a tranquilliser, and has been used during childbirth to produce the state known as 'twilight sleep', which is supposed to allow the pains of a difficult delivery to be forgotten. Doctors in mental institutions have injected it to control violent behaviour in their patients. Hyoscine was also used to treat other brain disorders such as the *delirium tremens* of alcoholism, and even the far more serious condition of meningitis. Sometimes it was prescribed as a sleeping pill, and it has been suggested that Crippen was attracted to using it in order to subdue his wife's sexual demands, which were no longer the ones he preferred.

Chemists can make the hyoscine molecule more soluble, or safer to take, by having a positive charge at the nitrogen atom. A quaternary nitrogen, in the form of the hydrobromide, increases the solubility in water, from which this compound can be crystallised as a white solid. Crystallisation has the effect of purifying the product and, almost as importantly, in the solid form it can be dispensed more accurately by weight. Hyoscine hydrobromide is a potent form of the drug which is absorbed through the stomach wall. As such it can be prescribed as a quick-acting sleeping pill.

Alternatively, a butyl group[ii] can be added at the quaternary nitrogen atom of hyoscine. In this case the drug is made safer, since

---

[ii] This is the *n*-butyl group, having the straight-chain structure: $CH_3–CH_2–CH_2–CH_2–$.

it can no longer pass across the **blood–brain** barrier, which in turn means that larger doses can be taken with less risk to the brain and central nervous system. This form of the drug is used to treat gastrointestinal disorders such as irritable bowel syndrome. The dose is 10 mg three times daily, although the dosage may be doubled if the condition is acute. Its most common use is to relax the smooth muscle of the intestine, the bile duct and the urethra.

Hyoscine has the curious effect of switching off that part of the brain which triggers vomiting, which is why it prevents travel sickness, especially sea-sickness. It is also used to treat the nausea and vomiting associated with chemotherapy. For this purpose hyoscine hydrobromide is the preferred form and the dose of 0.3 mg is repeated every six hours as necessary. Products containing hyoscine can be purchased over the counter under brand names such as Kwells or Joy-rides. For those undergoing chemotherapy a self-adhesive patch, Scopoderm, containing 1.5 mg of hyoscine, may be applied to the hairless skin behind the ear and will give relief for up to three days.

So how does hyoscine prevent seasickness? This condition is caused by a conflict of signals in the brain, some arising in the inner ear, which registers motion of the head, and others coming from the eyes, which are telling the brain that the body is not moving relative to its immediate surroundings. The ear has two organs containing fluid and these enable the body to maintain its balance. The brain coordinates this information with that from the eyes. For some people the conflicting information the brain is trying to resolve eventually activates the vomiting reflex. Chemotherapy can similarly cause the brain to receive mixed signals, especially when patients move their head, and this too can lead to nausea and vomiting. Why the brain responds this way is not understood, nor which specific receptors are targeted by hyoscine to prevent this happening. Some believe that it interferes with the message reaching the brain from the ear, whereas others think it acts to control the reflex action of the stomach muscles. It probably does both.

Other over the counter medicines containing hyoscine include Feminax and Buscopan, which are taken to counteract period pains. Feminax contains a small amount of hyoscine hydrobromide; Buscopan contains hyoscine butyl bromide, along with caffeine to counter the effect of drowsiness. Such products deliver a small dose of the drug and it is recommended they be taken at the onset of symptoms and at four-hourly intervals thereafter.

Recent research has revealed an unexpected benefit conferred by minute doses of hyoscine, and as a result we may one day see it used in the treatment of depression. In 2006 a report from the Mood and Anxiety Disorders Program at the National Institute of Mental Health, at Bethesda, Maryland, USA, reported a rapid reduction of depression in severe cases when hyoscine hydro-bromide was prescribed at a dose of 4 µg per kg body weight. A normal dosage would be around 0.3 mg (300 µg). The drug was injected as part of a **double-blind test** on a group of patients aged between 18 and 50, and the results showed this extremely low dosage to be very effective and quick acting.

2.4 Criminal Use

The criminal use of hyoscine relies on its power to reduce a person's inhibitions, their ability to process information, and their inability to remember what happened while under its influence. Hyoscine can cause a person to become talkative and delusional. It was in-vestigated by the Nazis and later by the CIA as a truth drug, but while it will indeed make a person highly suggestive to revealing information, even against their better judgement, the delusional aspects of hyoscine intoxication make the information unreliable.

Hyoscine is still used by criminals in parts of South America, especially in Colombia where it is known as burundanga. It can render a person incapable of resistance, making them almost like zombies, and thereby susceptible to theft or rape—or to providing access to their bank accounts. Memory loss favours the criminal, because the victims are subsequently unable to recall events or identify their assailants. In Bogotá there have even been instances of women being given hyoscine in order to steal their babies, the objective being to sell the child to an infertile couple who are desperate to purchase one for adoption.

The use of potions rich in hyoscine are said to have played a part in Colombian culture for centuries and there are tales of the wives and slaves of deceased Colombians chieftains being made to drink it before being entombed along with his body. It was said that under its influence they were willing to comply. The potion was probably derived from the leaves of the brugmansia tree. In Venezuela hyoscine is still extracted commercially from the leaves, stems and seeds of this tree to satisfy the world market for the drug.

The tree is known for its prolific trumpet-like white flowers with their heady odour. Another source of hyoscine is the *Duboisia* (corkwood) tree found mainly in Australia and New Caledonia. It too is grown commercially for its leaves which contain high levels of both hyoscine and hyoscyamine.

Somewhat unexpectedly, there is no officially accepted fatal dose reported for hyoscine when taken by mouth, but that did not deter the crime novelist Agatha Christie from plotting a murder using it in the play *Black Coffee*. This was her first attempt at playwriting and she had her victim, Sir Claude Amory, poisoned with hyoscine hydrobromide. The play was performed in London's West End in 1930 and later turned into a novel by Charles Osborne after Christie's death. There have been cases where individuals have consumed relatively large amounts and survived. In 1991 it was reported in the *Journal of Analytical Toxicology* that a 76-year-old man had made homemade wine from Angel's trumpet[iii] and then sampled it. Although he drank only a little of the wine he became seriously ill within the hour and was rushed to hospital suffering from breathing difficulty and general weakness. The wine was in fact a potent brew of hyoscine and it was estimated that he had imbibed more than 400 mg of the drug. He was lucky to survive. Angel's trumpet is also deadly on account of the **atropine** it contains, as we will discover in Chapter 5.

2.5 Dr Crippen Arrives on the Scene

Was Dr Harvey Crippen under the mistaken impression that a fatal dose of hyoscine could easily be taken by mouth? Maybe he had read that hyoscine was reputed to have been the favoured agent of poisoners in the Middle Ages. He certainly knew that it could kill, and from the severity of its symptoms he might well have assumed it had similar toxicity to arsenic, the preferred poison of domestic murderers at that time, of which a fatal oral dose was 4 grains (250 mg). This may explain why Crippen bought only 5 grains of hyoscine hydrobromide. A more likely reason, however, was that he chose it because he believed it to be almost undetectable, indeed the forensic evidence at his subsequent trial showed that it was indeed difficult to identify. (Arsenic on the other hand was easier to

---

[iii] Angel's trumpet is also known as moon flower and is prized for its beautiful and fragrant trumpet-shaped flowers.

detect and identify in the body after death.) Unfortunately for Crippen something went wrong with his plan. Had he overlooked the side effects of hyoscine? Did his wife begin to act like a woman possessed and begin to shriek and curse like a witch?

2.6   The Murder of Belle Elmore

Dr Hawley Harvey Crippen never confessed to murdering his wife, so we can only guess what happened on that fateful night when he gave her a large dose of hyoscine. Those who have attempted to find an answer have come up with various suggestions to explain his mysterious behaviour. He had obviously planned to kill Belle, but perhaps not in a manner which was quite so messy.

On 15 January 1910, Crippen ordered five grains of hyoscine hydrobromide from Lewis & Burrows Ltd, pharmacists of New Oxford Street, London, near where he worked. This was an unusual amount, but they knew Crippen as a regular customer who bought the types of poison used to make homeopathic remedies.[iv] Lewis & Burrows did not have enough hyoscine in stock so they ordered more from British Drug Houses, who in turn obtained their supplies from Merck of Germany. The cost was 9 pence per grain, equivalent to around £2.70 at today's prices. Crippen returned to the pharmacists on Wednesday, 19 January, to collect the hyoscine and signed the poisons register. The hyoscine was never seen again, although during his trial later that year he offered an explanation as to what he had used it for. He may have given Belle a small dose of the drug soon after he bought it, because she appeared to be somewhat disorientated a few days later at a meeting of the Music Hall Ladies' Guild, of which she was treasurer, and she spoke of waking one night with a sensation of being stifled.

It may have been that Crippen was hoping to poison Belle in a way that might suggest she was ill and that she had died from natural causes, for example that she had a heart attack and died in her sleep. It was the sort of thing which could happen to overweight people, and Belle was certainly a woman of ample figure. Whatever

---

[iv] Homeopathy operates on the belief that a poison or toxin which causes the same symptoms as a disease holds the key to its cure. However before a homeopathic medicine is offered to a patient it is diluted many times and shaken in a special way. The medicine is said to become more efficacious the more this dilution and shaking is performed. Indeed by the time the patient takes the medicine there may be not a single molecule of the original poison remaining in the solution.

his plan, Crippen decided that the fatal dose would be given on Monday, 31 January. That night he mixed a hot toddy for Belle, something he was known to do, and put the rest of the hyoscine in it. The idea was for her to become unconscious and die in her sleep, so that he could report finding her dead in bed the following morning; he would then telephone a doctor friend who would no doubt diagnose a heart attack. Crippen had already told the doctor the previous month that he was worried about his wife's health.

Clearly something went wrong with the plan, and it must have become clear that she was not going to die, despite a large dose of hyoscine. How he then killed her is not clear. Maybe he shot her with the revolver which was found when then police searched the house in July. Whatever happened, the result was a dead Belle whose body could not be passed off as having died a natural death—hence the need for Plan B. Crippen decided he had to get rid of her remains. His solution was to dissect her and bury her flesh under the coal cellar floor, where it would rot away in course of time. But his lack of chemical knowledge let him down, because he did something that would ensure her remains would not decompose. He bought a bag of **quicklime** (calcium oxide, CaO) and sprinkled it over them. This had the effect of dehydrating the flesh, and by reducing its water content preserved it from decay. As for Belle's bones, he may have burnt some of them or thrown them out with the refuse, or in the case of her head it has been suggested that he dropped it overboard when he was on a boat trip. Whatever he did with it, it was successful.

2.7   The Origins of Crippen and Belle

Crippen was born in 1862 in Coldwater, Michigan, the son of a dry goods merchant, and trained to be a doctor at the Michigan School of Homeopathic Medicine. During his student years he travelled to London, where he attended lectures at the leading hospitals, including the Bethlehem Royal Hospital, which specialised in mental disorders. There he witnessed how violent patients were sometimes given hyoscine hydrobromide and how this had a remarkable calming effect. On his return to the USA he continued his studies, graduating in 1884, and practised homeopathy in various American cities. Soon after graduating he married, and his wife Charlotte gave birth to a son, Hawley Otto, in 1890, but she died two years later in

January 1892 of "apoplexy", the term then used for a fatal stroke. At the time they were living in Salt Lake City. The baby was taken to California where it was raised by Crippen's parents.

Later that year, on 1 September 1892, Crippen married Corinne ('Cora') Mackamotzki in Jersey City. She was an aspiring singer and was already better known by her stage name, Belle Elmore. The following year she underwent an operation to remove her ovaries, which left her with a permanent scar. (This eventually enabled her remains to be identified.) In 1898 Crippen moved to London as the agent for his employer, Munyon's Homeopathic Remedy Company, and Belle joined him soon after. London offered lucrative opportunities for both of them. Belle hoped to exploit her obvious singing talents and maybe hit the big time in the music halls, while Crippen worked at Munyon's offices in New Oxford Street. In 1902 he was recalled to the United States for six months, while Belle remained in London. On his return to the UK, Crippen moved to the nearby Drouet Institute for the Deaf where he was guaranteed a minimum salary of £6 a week (£600, or $1200 at today's values) but he earned much more than this thanks to his involvement in a number of dubious mail order schemes. It was at Drouet's that he encountered a young typist, Ethel le Neve.

The Drouet Institute for the Deaf advertised a cure for deafness, and those who replied were sent literature including testimonials from satisfied customers, money-back guarantees, easy payment schemes and half-price offers. If they responded to the offer they were sent special plasters to put over their ears as the 'cure'. Needless to say, these did not work and the firm was eventually exposed by *Truth* magazine, which actually called one of the Drouet directors a quack, implying that he was not medically qualified. This was a libellous accusation, because he did have a medical degree, and it resulted in a court case during which Crippen gave evidence. In September 1909, Crippen moved to the Aural Remedies Company at Craven House, Kingsway, which dispensed another hearing cure, but it too was completely useless and was again exposed in *Truth* magazine, which actually linked Crippen to the previous scam at Drouet. He ignored the article.

By that autumn, however, he had other things on his mind, namely his desire to be free of Belle and to be able to marry Ethel le Neve, who had been his mistress for five years. Divorce in those days was out of the question for most married couples, so he decided to kill Belle by poisoning her. It would seem that he started to put his plan into action in December. He and Belle had a large cash deposit of £600 (equivalent to at least £60 000, or $120 000, in today's terms) earning interest at a fixed term rate with the Charing Cross Bank at a year's notice of withdrawal. That month he persuaded Belle to sign a withdrawal form. (In fact, the money was never reclaimed because the bank folded in the autumn of 1910, around the time of Crippen's trial.)

Belle was born in 1873, although at the time of the 1901 census she gave her age as 20, not 28. The census reveals that they were then living in an apartment in Bloomsbury, the bohemian district near Russell Square and only a few minutes walk from Crippen's office. Life in London may not have been all that Belle had dreamed of, but it was exciting. Her minor celebrity status ensured her entrée to the world of entertainment and the Crippens went to lots of dinners, parties and other social events. She obtained work in music halls around the country—although never making it to top of the bill—and she even appeared at prestige venues in London such as the nearby Holborn Empire. Her friends remembered her as a larger than life character with a taste for flamboyant clothes and expensive jewellery, some of which she paid for from her earnings and some of which Crippen purchased. On one occasion he bought her a designer coat costing £85 (equivalent now to £8500, or $17 000).

In 1905 Crippen and Belle moved to the up-and-coming suburb of Holloway to a house, 39 Hilldrop Crescent, which was to become world famous by being infamous.[v] The house was rented at £1 a week and had a basement which gave access to a back garden. It was meant to be a family home and had five bedrooms, two on the upper floor and three on the top floor. The house also had a toilet and bathroom. The Crippens sometimes took in lodgers, one of whom later said he often heard them quarrelling. The house was cluttered with Belle's costumes and was generally untidy, although they employed a daily maid, Rhoda Ray, who stayed late if the Crippens held a dinner party. Though the house appeared not to have been cleaned regularly, when they were entertaining guests they did it in style with the dining and sitting rooms always tastefully arranged. On 1 January 1910 they had wined and dined the

---

[v] The house was destroyed by a bomb on 8 September 1940 during the first major air raid of the London Blitz.

singer Lil Hawthorne, a famous music hall performer, as part of their New Year celebrations.

2.8   The Murder

The fateful evening of the murder was the last day of that month, and this time the guests for dinner were Paul and Clara Martinetti, who ran a troupe of clowns and acrobats. The invitation was rather sudden—Crippen called round at their apartment after work—but nevertheless they turned up at Hilldrop Crescent at 8 pm and spent what they were to recall later was a pleasant evening. The meal was followed by a few games of cards. They left for home at around 1 am, although they said it took Crippen a long time to find them a taxi cab. The suspicion later was that he had already slipped some hyoscine into Belle's drink and wanted the Marinettis to notice that she was feeling dizzy and unwell. The actual fatal dose was no doubt drunk by Belle soon after the Martinettis left, but something went wrong with Crippen's plan and Belle died in a manner which could not be explained as a natural death. Of necessity there had to be a sudden Plan B.

It appears that the following morning Crippen went to the office as usual and so must have only half completed his disposal of the body. He continued with his macabre work that evening, at the end of which he had a heap of flesh and a pile of bones, and also Belle's head. He lifted the bricks of the coal cellar floor and dug a shallow grave in which to bury the flesh. The bones he disposed of in a different way. Did he burn them? There were several fireplaces at No 39 where he could have done this, and then all he would have to do was break up the brittle remains and dispose of them in the ash bin for the dustmen to collect. Alternatively, he could have carried the bones out of the house in bags at night and discarded them, maybe throwing them into a canal. The so-called Holloway Sanitation Canal would have been the ideal place. Neighbours later remembered seeing Crippen working in the garden early in February, when he was no doubt disposing of soil from the grave in the cellar, and they also remembered him putting a lot of ash into his bin.

The day after Belle disappeared Crippen wrote two letters to the Music Hall Ladies' Guild, purporting these to have been dictated by his wife. One said that she had to return unexpectedly to the USA to see a relative who was seriously ill, and the second asked them to choose another treasurer as she did not know how long she would be away. That same day Crippen took two pieces of Belle's jewellery to a pawnbroker in Oxford Street and sold them for £80. A week later he sold seven more pieces for a further £115. The £195 these yielded would today be worth around £20 000 ($40 000). Their original purchase price would have been at least double this amount. As far as those who knew the couple were concerned, Belle was no longer around and Crippen said that she had gone back to her native country in answer to a telegram regarding a sick relative. America was a large country and it was most unlikely that anyone in London would go looking for her, either then or later after he had announced that she had died. Plan B appeared to be working.

Crippen made his first major blunder by showing Belle's large collection of jewellery to Ethel and asking her to choose some items for herself. She selected a particularly handsome star-shaped brooch which she wore at a variety artists' dinner, where Belle's friends instantly recognised it as one of her favourite pieces. Surely Belle would not have gone back to the USA and left that item in London? And why should Ethel le Neve, whom they knew was merely a typist at Crippen's workplace, be wearing it? They might turn a blind eye to Crippen having an affair, but not to his mistress's brass cheek in flaunting Belle's jewellery. What was going on?

On Sunday 20 March, Crippen wrote to the Martinettis mentioning that he had received a letter from Belle saying she was dangerously ill. Four days later they had a telegram from him to say that she had died at 6 pm on 23 March. Crippen placed an announcement of her death in the stage newspaper, *The Era*. That same week Crippen and Ethel le Neve went on holiday to France. As far as Crippen was concerned he was now a widower and therefore free to marry again. In France he hired an 18-year-old French maid. When they returned to London Ethel was wearing a wedding ring and she told her family and friends that she was now married. She said little about her new husband, and they were never to meet him. There had been no marriage, of course, but she had moved to Hilldrop Crescent, and was also wearing one of Belle's diamond rings as her engagement ring.

Who was Ethel le Neve? To the media she was the 'other woman' in the case. Ethel was the eldest of seven children and was born in January 1883 at Diss, Norfolk. Her family moved to London in 1890, and when Ethel was old enough to leave school her father sent her to Pitman's College to learn shorthand and typing. She then got a job at Drouet's, where she met Crippen and eventually they fell in love. He rented rooms at 82 Wells Street in the slightly disreputable area north of Oxford Street known as Fitzrovia, and there they slept together for the first time in December 1904. How many evenings they spent in their love nest we can never know, but no doubt while Belle was out of London, or even when she was on stage in the capital, it would not have been difficult for them to spend time together. Once they came close to disaster, when in September 1908 Ethel found herself pregnant, but she miscarried at an early stage and disposed of the foetus down the lavatory at her lodgings.[vi]

When Crippen announced that Belle had died in the USA, he said she had died of pneumonia while visiting his family in California. Some people sent letters of condolence to him or the Music Hall Ladies' Guild, and some even wrote to Crippen's son in California. The latter replied that his own son had recently died, but that he had no knowledge of his stepmother's death. The Music Hall ladies began to suspect that the story of her death was a lie and were amazed to hear that Crippen had remarried within a few weeks of Belle's death. Nothing came of their suspicions at this stage, no doubt in part because, in common with the rest of the nation, they were suddenly shocked by the death of the King Emperor, Edward VII, on 6 May.

Then in mid-June, two of Belle's celebrity friends, John Nash and his wife Lil Hawthorne, arrived back from New York and they too were puzzled to find that she was no longer around, and they could attest that they had not heard of her being in the USA while they were there on tour. They discovered that others in the entertainment world were equally puzzled by her sudden disappearance, and they went to Scotland Yard to report her missing and suggested that her disappearance was suspicious. They spoke to Detective Inspector Walter Drew, whom they already knew. He was sceptical about what they told him but he did agree to investigate further, and he called at Hilldrop Crescent on the morning of Friday, 8 July. There he interviewed Crippen, who admitted to Drew that the story of Belle's death was a fabrication designed to

---

[vi] Maybe with Crippen's help she took an abortifacient. He would almost certainly have known of drugs of this nature.

disguise the fact that she had walked out on him. Drew accompanied Crippen to his place of work and there took statements from him and from Ethel. He even went out for lunch with Crippen, who gave the impression of having nothing to hide.

Crippen suspected, however, that Drew was on to him. The following day Crippen sent his office assistant to buy some boys' clothes and shoes. Later that afternoon he and Ethel returned to No 39, where he cut Ethel's hair short and she put on the boys' clothes, which she found to be somewhat on the large size. They packed their suitcases and caught the train from Liverpool Street Station to Harwich, where they boarded the night boat to the Hook of Holland. From there they journeyed to Brussels and booked in the Hotel des Ardennes under the name of John Robinson and his 16-year-old deaf mute son, both supposedly from Canada.

Meanwhile back in London, Inspector Drew, along with a Sergeant Mitchell, returned to Crippen's office to question him further but when they arrived they were shown a letter left by him saying that he was going away. Drew immediately went round to No 39 where they made a forced entry and searched the premises, finding Crippen's revolver and evidence of his attempts to forge Belle's signature. He returned the next day to make a fuller inspection, but it was only on the day following that Sergeant Mitchell became suspicious of loose bricks on the coal cellar floor. When he prised some of them up he came across the mounds of human flesh. A warrant for Crippen's arrest was issued on 16 July.

When it came to burying Belle's remains, Crippen had again bungled things. Along with her flesh he buried parts torn from his own pyjamas and some hair to which curlers were still attached, and also pieces of female underwear. Clearly he had poisoned Belle after she had had time to prepare herself for the night. Belle's corpse must have presented Crippen with a problem: she was too heavy to carry by himself, and burying her in the back garden was not an option because this was overlooked by neighbouring houses—so he decided to bury her under the coal cellar floor. The hole he excavated for the purpose, however, was only big enough to bury her fleshy remains, and he had to dispose of her bones and head in other ways.

The forensic pathologist at Crippen's trial pointed out that whoever had done the dissection must obviously have had some previous experience, and Crippen had indeed dissected a corpse while at medical school. The shallow grave in which Crippen buried the fleshy parts was 4 ft long, 2 ft wide and 12 in deep, the earth from which he was able to dispose of unobtrusively around the garden. The quicklime he then sprinkled over the fleshy parts did not serve to speed up their decay by chemical attack, as he no doubt hoped, but rather to preserve many of the internal organs, including the heart, spleen, liver and intestines, and the stomach, which was almost empty.

2.9   Arrest and Trial

Drew issued a warrant for Crippen's arrest. The news of his sudden departure and the discovery of the remains at No 39 quickly become international news. Where were Dr Crippen and his lady friend? In those days passports were not needed in order to travel and it was obvious they might have left the country. They could now be anywhere, maybe on a boat sailing to a distant part of the far-flung British Empire. In fact they were in Belgium, in Brussels. They stayed there several days and eventually booked passage—a double cabin for £31—on the SS *Montrose*, departing from Antwerp for Quebec on 20 July. Its captain, Henry Kendall, who was already aware of the hunt for Crippen, soon became suspicious of his passenger, Mr Robinson, whose son looked rather effeminate, and when he saw the two of them squeezing hands in an obviously affectionate manner he suspected that they might well be the missing couple. He was still within radio contact with England, and he sent a signal saying that he was sure Dr Crippen and Ethel le Neve were on board. This use of the new-fangled radio caught the public fancy, as it was the first time it had been used in helping to catch someone fleeing from justice who was already at sea. Detective Inspector Drew immediately caught the train to Liverpool and there boarded the liner *Laurentic*, which he knew would reach Canada before the *Montrose*. The newspapers of the day reported the steady progress of the *Montrose* across the Atlantic, but those on board were oblivious to the drama that was unfolding and the fate which would soon await the fleeing lovebirds. On arrival in Canada, Drew joined the *Montrose* posing as a pilot and arrested the runaways. They were taken back to England and arrived in Liverpool on 20 August.

Crippen's trial began on Tuesday, 18 October 1910. Ethel was to be tried separately. It was the forensic evidence which condemned Crippen. It enabled the remains to be identified as Belle's, and the poison proved to be hyoscine. That the flesh was from Belle's body was confirmed by identifying the scar tissue which was the result of the operation to remove her ovaries. That piece of evidence was provided by Bernard Spilsbury, a brilliant newcomer, who was to become one of Britain's most influential and respected forensic experts. (The defence claimed that the mark on the skin was merely a fold.)

The pieces of men's pyjamas found with the remains were Crippen's, and were identified as those bought by Belle during the January sales in 1909. They were made from material that had only become available two months earlier, thereby proving that the grave under the cellar floor was recent. Among its contents were curlers around which the hair was still wound, and this was long and bleached and again consistent with their being from Belle. The purchase of the hyoscine also told against Crippen, although he claimed he had used it to make 'pillules' for treating nervous diseases. No evidence for such pills ever came to light, nor was anyone found who had bought them. In fact hyoscine was never a component of homeopathic remedies.

The presence of hyoscine in the remains was proved by Dr William Henry Willcox, a Home Office forensic scientist. He extracted one thirtieth of a grain (2 mg) of hyoscine from the stomach, one fortieth (1.6 mg) from the kidneys, a twelfth (5.5 mg) from the liver, and a seventh (9 mg) from the intestines. This distribution of hyoscine was consistent with its having been taken orally, rather than injected, and that it was taken a few hours before death. The total amount extracted was two-sevenths of a grain (18.6 mg) and Willcox deduced that the body had contained at least half a grain (32 mg), which he said implied that Belle had been given a fatal dose.

"How can you be certain it was hyoscine?" asked the prosecution lawyer. Wilcox carefully explained to the court the tests he had carried out. He had tested the remains for mineral poisons and plant toxins. He had detected small amounts of arsenic but at levels that were not significant—indeed this is a natural component of the human body. He discovered that the poison was a plant toxin of the alkaloid type, and he confirmed this by putting a drop of solution extracted from the flesh into the eye of the hospital cat of

St Mary's Hospital where he lectured. The effect was to dilate the pupil, an infallible test for this kind of poison. He then tested for common alkaloids such as morphine, strychnine and cocaine, but none of these was detected. Willcox also had to exclude human decomposition products from his test. Eventually he became certain that the alkaloid was either atropine, hyoscyamine or hyoscine. Until quite recently it had been impossible to differentiate these by chemical analysis but Willcox was finally able to prove that it was hyoscine. He told the court that patients given hyoscine quickly become drowsy, would be unconscious within the hour, and likely to die within 12 hours. A Dr Luff, a science adviser to the Home Office, was also called to the witness stand and he supported Willcox's findings. (The defence called a witness who said that the alkaloids found by Willcox might well be decomposition products from the body, but this evidence was not accepted.)

In 1898 the law had been changed to allow an accused person to give evidence in court. Crippen chose to do so, but he was unable to answer the prosecution's questions as to why he had made no attempt to find Belle if he believed she had left him. He also found it hard to explain what had become of the hyoscine he had purchased. He said he had dissolved it in alcohol and used the solution to make the pillules, each of which he estimated would have contained only 1/3600 of a grain (0.02 mg). He said he had supplied these to clients, although he could produce none of them to support this.

On the Saturday afternoon, 22 October, the all-male jury took only 27 minutes to find Crippen guilty. The Lord Chief Justice, Lord Alverstone, delivered the sentence: "Hawley Harvey Crippen, you have been convicted, upon evidence which could leave no doubt in the minds of any reasonable man, that you cruelly poisoned your wife, that you concealed your crime, you mutilated her body, and disposed piece-meal of her remains. You possessed yourself of her property and used it for your own purposes. It was further established that as soon as suspicion was aroused you fled from justice and took every measure to conceal your flight. On the ghastly and wicked nature of the crime I will not dwell . . . I have now to pass upon you the sentence of the court, which is that you be taken from hence to a lawful prison and from thence to a place of execution, and that you be hanged by the neck until you are dead . . ." His execution was set for 8 November but there was an appeal, on the basis that the evidence about the pyjamas was given after the defence case had been made, and in any case the remains had not been proved to be from a female. But it was a vain hope, and the Court of Appeal rejected these arguments. The appeal was heard on 5 November, and this necessitated the execution to be put off until the 23rd of that month.

Modern forensics would have treated the scene of crime very differently and been able to prove that the remains were definitely those of Belle from their DNA—see below. Investigation of the house would have revealed exactly where Crippen had dissected the body: either in the bath, as some assumed, or on the basement floor next to the kitchen. Even the theory that he might first have drugged her with hyoscine and then shot her could have been proved by inspecting the gun and finding telltale blood spots in the room in which it had been fired—if this was indeed how it had happened.

Ethel le Neve's trial was completed in a day. It started on the morning of 25 October and it was over by 4:15 pm, with Ethel walking free. Soon after Crippen had been hanged, Ethel sold her story to *Lloyd's Weekly News* and the *Daily Chronicle*. In 1915 she married Stanley Smith, who worked in a furniture store, and they had two children. Thereafter she faded from public view and led a life of quiet domesticity, protected in part by her new name. She died of heart failure in August 1967, aged 84.

Could Crippen have escaped his fate? Probably. His choice of poison might well have allowed him to claim that Belle had died of natural causes, and if so an autopsy would not have been carried out with the care that Willcox used. The hyoscine would not have been discovered, let alone identified. Clearly something happened on that fateful night to make him resort to killing Belle in a way that could not be disguised as a natural death. Even Crippen's Plan B might have succeeded had he not used lime to preserve Belle's remains. And although his encounter with Inspector Drew had been a success in allaying the detective's suspicions, as Drew later admitted, by fleeing the country Crippen showed he had something to hide.

As far as Dr Crippen was concerned the devils that brought him to the hangman's rope were all in the detail, and especially the chemical ones.

2.10 Tailpiece

Or has he been misjudged all this time? According to noted toxicologist John Trestrail III, of Grand Rapids, Michigan, the remains found under the cellar floor of 39 Hilldrop Crescent were not those of Belle. In October 2007 he claimed to have DNA evidence that proved this remarkable claim. Trestrail enlisted the help of David Foran of Michigan State University and genealogist Beth Wills and together they tracked down living descendents of Belle's sister and compared their mitochondrial DNA with that on a microscope slide from the Royal London Hospital archive. The slide was part of the evidence gathered at the time of the trial, and the specimen of tissue had been preserved in formaldehyde and the slide sealed with an adhesive. Unfortunately the glass slide had to be broken in order to extract tissue for analysis. What Foran discovered was that there was no match between its DNA and that taken from Belle's grand-nieces. Trestrail could only conclude that the tissue from the slide was not from Belle, in which case Crippen was not guilty of her murder—although he certainly had killed someone. Was it a woman on whom he had performed an illegal abortion which went wrong? That is what Trestrail suggests—but in that case who was she? And what became of Belle?

This new evidence conflicts with all the other forensic evidence and one can only assume that either the sample of tissue on the slide was not that of Belle, or there is no direct genetic link between her and those living today. Both seem unlikely, and yet the DNA evidence is difficult to refute. No doubt the Crippen story will continue to run and run.

CHAPTER 3

Atropine and Mrs Agutter's Gin and Tonic

*A word shown in **bold** indicates that further information will be found in the Glossary*

3.1 "Deadly Nightshade, Dear?"

Atropine is chemically similar to hyoscine, the subject of the previous chapter. Indeed the two differ by a single oxygen atom, a simple change which makes a great deal of difference to their behaviour in the body. Atropine is far more poisonous than hyoscine. A single berry of deadly nightshade contains enough atropine to kill a young child, and yet there are people who have survived high doses of this toxin. Alexandra Agutter was not meant to survive, but she did. Her husband tried to murder her with atropine by disguising its bitter taste in a gin and tonic. He then tried to throw the blame on to some mysterious psychopath who had spiked several bottles of tonic water with atropine and replaced them on the shelves of an Edinburgh supermarket. Agutter's scheme reads like the script of a crime investigation programme and, as in some of these, he was brought to justice thanks to forensic chemistry. Like the villain in such programmes he too ended up in jail, but unlike them he subsequently found a job lecturing students on philosophy and ethics at Manchester University. Real life is often stranger than fiction.

Molecules of Murder: Criminal Molecules and Classic Cases
By John Emsley
© John Emsley, 2008

3.2 Atropine

Atropine is a mixture of two **chiral** molecules and is chemically DL-hyoscyamine. The older name, atropine, is related to one of the three Fates of Greek legend. They were Clotho, Lachesis and Atropos, and they visited every newborn child and determined what its future would be. Clotho spun the thread of life from light and dark fibres, Lachesis twisted the thread, sometimes making it stronger, sometimes weaker, and Atropos held the shears which finally cut it off. She determined how long a person would live, and so it must have seemed appropriate to name a genus of poisonous plants *Atropa* after her, and the deadly chemical which they produced became known as atropine. Strange at it appears, atropine can also save lives, revealing a side to its nature that conflicts with the origin of its name, as we shall later see.

Atropine is a white, odourless, crystalline powder which was first isolated in 1833 by two German chemists, Philipp Lounz Geiger (1785–1836) of Heidelberg and his student Germain Henri Hess (1802–1850).[i] They extracted it from the black, shiny, cherry-sized berries of the deadly nightshade (*Atropa belladonna*) and also from its leaves. Belladonna, as the plant is more commonly called, is a native of woodland and waste areas in central and southern Europe and is still grown commercially in France as a source of the atropine used in sedatives, stimulants and antispasmodics. Atropine is also stockpiled for military purposes—not as an offensive poison but as a defence against chemical warfare agents. The demand for atropine has led to the selective breeding and genetic modification of belladonna to increase the yield. Belladonna is a shrub which grows to about 1.5 m (4–5 ft) and has dark green leaves and violet flowers. Its fruit looks tempting enough to eat, but anyone who unthinkingly chews a berry will soon spit it out because it is very bitter. For this reason very few people die accidentally of atropine poisoning. The related black nightshade (*Solanum nigrum*) is also poisonous while its berries are green and unripe, but once they become black and fully ripe they can be cooked and eaten without ill effect.

The name belladonna comes from the Italian *bella donna*, meaning beautiful lady, and the earliest use of the term was in 1554

---

[i]There is some evidence that an apothecary named Mein may have extracted it earlier, in 1831.

by Pietro Andrea Mattioli (1501–77), a distinguished physician living at Sienna in Italy. He reported that actresses and courtesans in Venice would squeeze a drop of the juice of a nightshade berry into their eyes. The resulting dilation of the pupils produced a fashionable doe-eyed appearance and one treatment would last for several days. Using atropine in this way became widely popular in the 1600s and 1700s, not only among actresses but also among ladies of fashion. The procedure was not entirely without risk, however, because frequent use of atropine eye-drops can lead to long-term damage to the eye and even to blindness. While European women used atropine for cosmetic purposes, people in other parts of the world found very different uses for belladonna. In Morocco it was considered to be an aphrodisiac, in Nepal it was regarded as a sedative, and in the Middle East it was believed to reinforce the effects of marijuana.

American teenagers revived an interest in the drug in the 1980s and experimented with a drink made from the leaves of the angel's trumpet bush, which are also rich in atropine. The aim was to experience the curious hallucinations which this natural chemical can produce once it crosses the **blood–brain barrier**, but this can be dangerous. Too much atropine can make a person seriously ill, and the US Center for Disease Control has issued warnings about the dangers of angel's trumpet. Such was the concern about its misuse in Maitland, Florida, that the city council even made it illegal to plant this ornamental bush in gardens.

3.3 Physiological Properties

The atropine used in medicine is a 50 : 50 mixture of the left-handed (L) and right-handed (D) stereoisomers of the chemical **hyoscyamine**, and in medical texts it is therefore written DL-hyoscyamine. Plants produce the compound as the L-isomer but during its extraction it converts to a mixture of the two **chiral** forms. This naturally occurring drug is the active agent of the herbal remedy known as belladonna herb, which was sold at one time as finely powdered dried leaves as a traditional remedy for intestinal colic, and given with laxatives to prevent gripes. The hyoscyamine content was about 0.3%, and doses of between half a grain and three grains (33–200 mg) were prescribed. Belladonna root was also used, and this contains slightly more atropine, around 0.4%. It was added to ointments for the relief of pain, especially that from haemorrhoids (piles), and was also available as a suppository.

Sometime the dried leaves of plants rich in hyoscyamine have found their way into the wrong product. This happened with so-called Paraguay tea, which is made from the dried leaves of *Ilex paraguariensis*, a holly-like bush. These contain **caffeine** and the chemically related **theophylline**, which act as mild stimulants.[ii] In March 1994 a batch of Paraguay tea contaminated with dried belladonna leaves was imported by a New York herbalist, but thankfully very little was sold before the error was noticed. However, a 30-year-old man and his 38-year-old wife did buy some and drank the tea made from it, and within half an hour they both felt so ill that they rang the emergency services and were rushed to hospital. The man, who was now seriously ill, had to be injected with the antidote, physostigmine. His wife needed less drastic treatment and fortunately both soon recovered. The following day a 20-year-old woman also drank some tea made from the contaminated material and was similarly admitted to hospital and her stomach pumped out. Two days later a family of four were afflicted. The New York City Department of Health were called in to investigate and they tracked down and impounded the Paraguay tea, analysis of which confirmed it to be contaminated with belladonna leaves. In the USA there are around 500 cases of accidental atropine poisoning each year, of which only a few individuals show symptoms severe enough to require hospital treatment. The source of the atropine is invariably belladonna and usually results from atropine-containing plants being incorporated into herbal products.

Atropine is not very soluble in water, but its solubility can be improved by reacting it with sulfuric acid to form atropine sulfate, which is available under such trade names as Atrophate and Atropisol. Alternatively, and to reduce its toxicity, a methyl group can be added at the nitrogen atom. This compound is used medically as atropine methyl bromide (Tropin) and the more commonly used atropine methyl nitrate (Atrodote, Eumydrin, *etc.*). These have less effect on the central nervous system and are less toxic than atropine itself or its sulfate salt. In the form of its

---

[ii] Theophylline has one fewer methyl group than caffeine.

salts the drug is more soluble in water, the sulfate remarkably so: 100 ml of water can dissolve more than 100 g of the sulfate, compared with only 0.25 g of atropine itself. A mere 1 ml of water is therefore able to dissolve many times the fatal dose of atropine sulfate, which is around 10 mg for children and 100 mg for adults. A single drop of such a solution is thus a very potent murder weapon—at least in theory. Its drawbacks are twofold: firstly, it has an acutely bitter taste which alerts a potential victim that something is wrong, and secondly its symptoms are easily recognised, and treatment is almost always effective. We shall see that both the warning and the successful treatment played their part in the attempted murder of Alexandra Agutter in 1994.

There is a marked difference in people's susceptibility to atropine, and some adults have survived doses as high as 1000 mg (1 g). It is even possible to suffer atropine poisoning by eating the meat of cattle and rabbits which have grazed on deadly nightshade, although the animals themselves do not appear to be affected. The honey collected by bees from the flowers of this plant has also given rise to cases of atropine poisoning.

3.4 Signs and Symptoms of Poisoning

The symptoms of atropine poisoning were once summarised in the following way: hot as a hare, blind as a bat, dry as a bone, red as a beetroot, mad as a hatter. 'Hot as a hare', because the body temperature goes up and stays up for as long as 8 hours; 'blind as a bat', since the pupil of the eye is so dilated that everything appears blurred; 'dry as a bone', due to suppression of the salivary glands, causing total dryness of the mouth, severe thirst and difficulty in swallowing; 'red as a beetroot', due to dilation of the blood vessels of the skin, so much so that atropine poisoning was sometimes misdiagnosed as scarlet fever (scarlatina); and 'mad as a hatter' on account of the rambling speech and wobbly gait it causes, resembling a person who is drunk.

A less dramatic description of the symptoms of atropine poisoning includes thumping of the heart, laboured breathing, feverishness, excitability and hallucinations. Some people also suffer a rash on the face, neck and upper trunk. Another symptom of atropine poisoning is that it paralyses the bladder muscles, causing urine retention which can be relieved only by inserting a catheter. The usual symptoms of a low dose of the drug are intense thirst, sensitivity to bright light, unsteady gait and mental confusion.

Atropine has its most deadly effect on the autonomic nerves, those which govern vital muscles over which a person has no direct control, such as those of the heart, lungs and gut. It also affects the central nervous system, causing mild disturbance, and higher doses produce agitation and disorientation. Some people are hypersensitive to atropine and find that even tiny doses can cause conjunctivitis of the eye and a rash on the body, indeed some patients who have been prescribed atropine have ended up with severe eczema.

Atropine is quick to act, particularly when injected, and its effects are felt within the first minute. Even when ingested, it begins to work soon after reaching the stomach, since it is readily absorbed into the blood stream and within 15 minutes its symptoms start to become apparent. Studies using atropine labelled with radioactive carbon have shown a peak of activity, as recorded by heartbeat, 15 minutes after an injection. This research was performed on four volunteers who were each given a dose of 2 mg. Most of the radioactivity, between 75 and 95%, was found to be excreted in the urine, and about half of this was unchanged atropine. It was also possible to detect traces of radioactive carbon dioxide in the exhaled breath. Within the body atropine has a half-life of two hours, but it can take three to four days for all traces to be completely removed from the system, and some of its effects can remain for as long as two weeks. About one person in ten who consumes a potentially fatal dose of atropine is liable to die, but with prompt treatment this outcome is unlikely.

Atropine, like hyoscine, interferes with the ability of the messenger molecule acetylcholine (ACh) to pass on signals from nerve endings to the receptors which are being activated. Atropine does this by blocking the receptors, particularly those in the smooth muscle[iii] and the brain, and its effects are somewhat different to those seen with hyoscine. Atropine inhibits the action of the glands which control the release of saliva, tears, phlegm and sweat, and

---

[iii] Smooth muscle tissue constitutes most of the muscles of internal organs of the body, including the digestive system.

this occurs at very low doses. Larger doses noticeably affect the eye, causing the pupil to dilate, and if the dose is high enough the eye muscle becomes paralysed. (Dilation also results from the direct application of atropine eye-drops, and this effect was at one time employed by ophthalmic surgeons when examining the eye, although a safer drug, tropocamide, is now preferred.)

Atropine affects both the central nervous system and the peripheral nerves, which it first stimulates and then damps down. It has been used as a treatment for the spasms associated with conditions such as Parkinsonism, and it can improve the walking, sitting and the speech of those affected. At one time atropine cream was applied to relieve the pain of rheumatism, sciatica and neuralgia, although there was no scientific basis for this. Plasters containing atropine were widely used in the 1800s, and as recently as 1950 they could be obtained over the counter as treatments for a variety of conditions including tuberculosis and mastitis, although they probably offered little benefit. Unlike hyoscine, atropine is in fact not easily absorbed through the skin, but it can gain access through open wounds.

Atropine is an **anticholinergic** and is therefore prescribed as an antispasmodic to relax the muscle wall of the intestines, and as such has been given to relieve stomach cramps and irritable bowel syndrome. The amounts of atropine that are given for therapeutic purposes have to be tiny, typically less than 1 mg. Atropine is still used as a premedication before anaesthesia in order to reduce salivary and bronchial secretions, preventing these from blocking the airways. At one time atropine was prescribed for the treatment of conditions in which there was excessive production of bodily fluids, such as hay-fever, colds and diarrhoea. It was even prescribed as a cure for bed-wetting. Doctors have also been known to give atropine to suppress the laboured breathing (death rattle) of a dying patient if this become distressing for those keeping a bedside vigil. Because of its undesirable side effects, atropine is less used nowadays and safer modern drugs are available which are more effective.

3.5   Treatment of Atropine Poisoning

Atropine poisoning is dealt with by first pumping out the contents of the stomach and then flushing with copious amounts of water.

A purgative, such as a strong solution of sodium sulfate, is given to stimulate the wave-like motion of the intestines and expel their contents. Diazepam may be given simultaneously to suppress convulsions. In 1969 four patients were mistakenly given doses of 500 *milli*grams of atropine in place of the prescribed 500 *micro*grams, and quickly passed out. The patients were immediately given injections of physostigmine, a drug extracted from Calabar beans, which is known to have the exact opposite effect to atropine, in other words it stimulates the receptors which are switched off by atropine. The patients recovered after about 8 hours and suffered no further adverse effects. (Phytostygmine by itself is a deadly poison, and atropine is injected as its antidote.)

3.6   History of Atropine as a Poison

One of the earliest references to atropine in the English language was by a mediaeval apothecary called Gerard who warned of the dangers of deadly nightshade, which he called 'sleepy' nightshade and which he knew could be fatally poisonous:

> " . . . this kind of Nightshade causeth sleepe . . . it bringeth such as have eaten thereof into a ded sleepe wherein many have died."

Ancient civilisations already knew of its dangers, indeed atropine has a long history of use and misuse. The ancient Greeks were aware of the effects which the juice of deadly nightshade berries could produce. The plant was associated with the cult of Dionysius, whose worshippers engaged in secret revelries in which wine doped with the juice was drunk and incredible visions were experienced. (The hallucinations associated with atropine were often described in terms of diaphanous fabrics.) The Greek physician Dioscorides warned of the dangers of deadly nightshade, which he knew to be far more toxic than mandrake (hyoscine). Deadly nightshade was believed to have occult powers and down the ages many folk hung a sprig of it in their homes to ward off evil spirits.

One of the most famous, indeed infamous, poisonings associated with belladonna was that ordered by Cleopatra. She decided to commit suicide by poisoning herself after the naval battle of Actium, fought off the coast of Greece in 31 BC, in which her lover

Mark Antony's fleet was defeated by that of Octavian, who eventually became the Emperor Augustus. She wanted death to be quick and painless and thought that the juice of the deadly nightshade might serve her purpose, so she ordered a slave to drink some, and observed that while he died a quick death it was clearly not a painless one. Further toxicological research was needed—none of which involved animals, of course—until she was finally persuaded that a bite from an asp was just as quick and offered a much more tranquil end to life, so she chose that—or so it was said.

This version of Cleopatra's suicide was in fact written more than a century after the event and as a result is not altogether reliable. It is much more likely that she was murdered. This is the theory proposed by Pat Brown, a leading criminal profiler in the USA, who put forward an alternative and more convincing explanation in a programme on the History Channel in 2004 called *The Mysterious Death of Cleopatra*. Brown concluded that Cleopatra was assassinated for political reasons by agents of Octavian—as was her son by an earlier Roman lover, Julius Caesar—and pointed out that Octavian had motive, means and opportunity. The story of the asp and the suicide note, though far more dramatic, was simply untrue.

However, Cleopatra may have been familiar with the juice of the belladonna berry. A recently discovered coin dating from 32 BC has Cleopatra's head on one side and suggests that perhaps she was not quite the beauty she was thought to be. How then did she fascinate both Julius Caesar and Mark Antony, the two most powerful men of their day, to the extent of having love affairs with each of them? Her beauty has always been assumed to have been an irresistible feature, and down the ages she has been portrayed in plays as a stunningly good-looking woman ("Age cannot wither her . . .", Shakespeare, 1606), in poems ("And she was fair as is the rose in May . . .", Chaucer, 1382), and in films. (Elizabeth Taylor as Cleopatra in the 1963 epic, *Antony and Cleopatra*.) The coin, which came to light in Newcastle-upon-Tyne in the north-east of England, depicts a woman with a pointed nose and chin and a rather thick neck. What it also shows are slightly bulging eyes. Cleopatra may have used atropine to dilate her pupils, knowing that this gives an alluring signal to members of the opposite sex. It is well known that our pupils dilate when we are particularly interested in another person, and that this message is subconsciously picked up by the intended recipient. Perhaps she had overdone the amount of belladonna juice she put in her eyes, and thereby caused them to swell.

Belladonna juice was widely known in Rome to be a deadly poison and as such was often misused. The problem was one of long standing and in 82 BC the Romans passed a law to suppress domestic poisoning, which appeared to be an endemic problem. It was said that there were professional women poisoners who were prepared to dispense belladonna to clients eager to dispose of people they no longer wanted around. The law appears to have had little effect, and mysterious deaths among the upper classes continued to happen. Juvenal, who wrote satires of Roman life in the period 100–127 AD, was critical of much that he saw around him and some of his phrases have come down to us, his most famous being the cynical remark that 'all we need from those who rule us are bread and circuses'. However, he does mention the deadly effect of atropine and that the poison came from deadly nightshade—and that it was favoured by wives wishing to get rid of unwanted husbands, even husbands who were emperors, or so it was whispered.

Rumour has it that Livia, wife of the great Augustus, Emperor from 27 BC to 14 AD, murdered her husband using a fig containing atropine. The emperor apparently had a favourite fig tree, whose fruit he particularly liked and which was reserved exclusively for him. Livia is said to have squeezed the juice of some belladonna berries into one of the figs. Whether this was true we can never be sure, but the circumstantial evidence seems to indicate that those around her were liable to die in a manner that suggested she had a hidden agenda, namely to ensure that her son would one day be emperor.

Livia Drusilla (58 BC–29 AD) was born into an upper class Roman family and was married to a cousin of high rank. The marriage took place when Livia was 16 years old, but unfortunately her husband backed Mark Antony in the civil war which was then raging and, after Antony lost the decisive battle of Philippi, the couple had to leave Rome and live in Greece until things quietened down. After a general amnesty had been declared they returned to the capital and it was then that Livia first met Octavian, the victor in the civil war, now totally in control as the Emperor Augustus. It appears to have been love at fist sight. Within months the emperor had divorced his own wife, who actually gave birth to a daughter, Julia, on the day of the divorce. Livia also was heavily pregnant by her husband, but he agreed to an immediate divorce. Three days after her baby was born she married Augustus. That marriage lasted 51 years but produced no direct heir for him.

Livia had two sons, Drusus and Tiberius. The former became a general in the army and the latter the next emperor in 14 AD after being legally adopted by Augustus. Did Livia scheme to bring this about? The Roman writers Tacitus and Dio Cassius clearly thought so, but they were writing long after the event, two centuries later in the case of Dio Cassius. Nevertheless it was strange that so many young men who might have claimed the imperial throne died prematurely. Julia, Augustus's daughter, had married and produced three sons, any of whom could have been expected to succeed Augustus. They were Lucius, Gaius and Agrippa Postumus, the last of whom had earlier been adopted by Augustus as his own son. They all died young.

No doubt other politically motivated murders have been carried out down the centuries using belladonna juice, although arsenic became the preferred agent in the late Middle Ages. Despite the infamy and known dangers of deadly nightshade it was nevertheless included in the pharmacopoeia in the 1700s and 1800s. In the Edinburgh Dispensatory of 1803 it was said to be efficacious for treating fevers (and particularly good against the plague), palsy, apoplexy, epilepsy, whooping cough, hydrophobia (a characteristic of rabies), and mania. It was to be taken as the powdered root, and if that did not work, as an infusion made with hot water, the dose being increased day by day until a 'tension of the throat' developed, after which the dose ought not to be increased further. By this time it was already known that extracts of belladonna affected the eye, and a Professor Reimarus advocated using it prior to removing a cataract.

In 1911, R A Witthaus reported in his *Manual of Toxicology* that he was aware of 682 clinical cases of atropine poisoning, of which half were caused by belladonna eye drops. More than 600 of the cases were accidental, but 37 were suicide attempts and 14 were murders. Of the total, 60 had died. Today they would probably all have survived but back in 1911 the facilities for intensive care did not exist.

3.7 Atropine as Antidote

It seems improbable that a deadly poison such as atropine could act as an antidote but such is the case, indeed it can offer protection against some of the deadliest poisons ever devised—the nerve gases. Atropine has also long been known as an antidote for another toxin, physostigmine. The action of atropine in blocking physostigmine was discovered by Thomas Fraser (1841–1920), Professor of Pharmacology at Edinburgh University, whose research was a landmark in the development of pharmacology. Physostigmine acts on the vagus nerve, which reaches from the lower end of the brain to the heart, lungs and stomach, causing the heart to slow and the stomach to contract, but these effects can be countered by the action of atropine.

The action of atropine as an antidote again involves the messenger molecule, ACh. When this has completed its task of passing a signal across the synapse gap, the body needs to remove it and does so using enzymes known as cholinesterases. These enzymes can be rendered inactive by a variety of chemicals, in particular by **organophosphorus** compounds and including the organophosphates. Organophosphates are used as insecticides. Organophosphorus compounds, on the other hand, are in general much more deadly and have been deliberately developed to be so. These are the so-called nerve gases, of which the best known is sarin. Sarin is absorbed by the cholinesterase enzymes, which are then put out of action. Meanwhile ACh continues to be generated by the nerves, building up in the synapse gap and repeatedly triggering the same receptors, causing the organ concerned to malfunction by going into spasm and finally becoming paralysed. Nerve gases block the enzyme so effectively that one tiny droplet represents a lethal dose. One of the first symptoms of sarin poisoning is semi-blindness, but it then quickly kills by paralysing the nerves and muscles of the lungs and heart. Sarin is not actually a gas but a colourless liquid boiling at 147 °C, yet it is volatile enough for its vapour to contaminate the air at lethal levels. Its victims do not realise they are in danger, because sarin provides no warning odour.

Sarin poisoning can be treated by the simultaneous rapid injection of two antidotes: atropine and pralidoxime. Atropine is the more important because it blocks the receptors which ACh is targeting, and this checks the damage their over-stimulation causes. Then pralidoxime is able to destroy the sarin molecules and so free up the cholinesterase enzymes. Soldiers in war zones are issued with syringes containing atropine and pralidoxime with which to inject themselves in the event of a nerve gas attack. On the other hand, soldiers who inject themselves with atropine but have not been attacked by a nerve gas are then at risk of dying of atropine poisoning. This may have happened to some US soldiers during the Desert Storm phase of the Gulf War of 1990–91, when some of them thought they were under attack from chemical warfare agents.

Nazi scientists developed sarin during World War II, and by 1944 it was being produced and stockpiled in bulk. It was not deployed, because Hitler believed that allied scientists must also be in possession of this deadly weapon and that they would retaliate with it on a larger scale. This may have deterred Hitler, but others have been less reticent about using it. In 1988, Kurdish villagers were the victims of a gas attack by the Iraqi government of Saddam Hussein which left many women and children dead. The Iraqis denied that a nerve gas had been used and at the time it was impossible to prove it, but it has since been confirmed that the Kurdish accusations were true. Shrapnel collected from bomb craters in Kurdish villages in 1993 was smuggled out of Iraq and taken to the Chemical and Biological Defence Establishment at Porton Down, UK, where chemists were able to detect minute traces of sarin absorbed by the paint on a fragment of shrapnel.

A second sarin attack against unprotected civilians took place in Tokyo on 19 April 1995. Twelve people died and more than 5000 were injured when members of the Aum Shinri Kyo doomsday sect released the agent on the Tokyo underground during the morning rush hour. An earlier release of sarin by the same group in 1994 had killed seven people in Matsumoto, Japan. That incident had alerted the authorities to the terrorist threat and consequently when passengers on the Tokyo underground began to collapse their symptoms were soon identified as being due to sarin. Many lives were saved because the atropine antidotes were on hand, and thousands of affected commuters were able to return home the same day. The most common symptoms were blurred vision, a throbbing headache and laboured breathing. For most of the victims the damage to the eyes recovered only slowly and was particularly painful. Shoko Asahara, the leader of the religious cult who inspired the attack, was arrested, tried and jailed.

3.8 Atropine Murders in Fiction and in Fact

The American popular author, Jack London (1876–1916), who wrote more than 40 books of which *The Call of the Wild* was the most successful, is thought to have committed suicide using an injection of atropine. He also gave himself an injection of morphine at the same time, perhaps realising from the Cleopatra story that death from atropine by itself would be painful. There is some doubt about the cause of London's death because he was suffering from a serious kidney disease, and it has been suggested that he might have died of an accidental overdose of morphine, in which case atropine was not involved.

Atropine has been a factor in murder mysteries. The Canadian crime novelist, Charlotte MacLeod (1922–2005), was very much aware of its effects and in one of her books, *Something in the Water,* published in 1996, the murderer employed atropine in eye drops.

The serial poisoner, Graham Young, experimented with atropine in November 1961 when he was only 14 years old. He put some in his sister Winifred's early morning cup of tea. Because of its bitter taste she drank only part of it, but it was enough to cause her to become giddy as she journeyed to her work at a music publisher in central London. When she arrived it was clear that something was wrong and her employer sent her immediately to the nearby Middlesex Hospital, where atropine poisoning was diagnosed. When she confronted her brother that evening he denied having tampered with her drink and tried to persuade her that the chemicals in her shampoo could have caused her symptoms.

Later Young moved on to other poisons, and in particular to thallium, with which he murdered his stepmother, a crime to which he confessed and was sent to Broadmoor, a secure hospital for the criminally insane. He appeared to respond to treatment and was released in February 1971. He obtained a job at a factory making lenses and then began to poison his workmates with thallium, some of whom died slow and painful deaths. Thallium is an insidious poison with a time delay of several days before it begins to work, a valuable feature for assassins (Chapter 10).

If a murderer intends to kill someone with atropine they will need to be given at least 100 mg and delay the arrival of medical help if it is to be reasonably certain that the person will die. What might attract a murderer to using this poison is the fact that it is almost impossible to detect at autopsy, and if the corpse is buried the toxin will disappear rapidly. Atropine distributes itself to all parts of the body and there are no inflamed internal organs for a pathologist to discover when conducting a post-mortem examination.

Atropine in the form of belladonna drops was on one occasion used by a murderer in an attempt to disguise the effects of a massive done of morphine which he had administered to his wife in order to inherit her wealth. (Murder by morphine is the topic of the next chapter.)

3.9 The Murder of Anna Buchanan

Anna Buchanan was the second wife of Dr Robert Buchanan; he murdered her in New York City in 1892. Buchanan was by birth a Canadian from Nova Scotia. He studied medicine in Edinburgh and set up in practice in New York in 1886. Although he was married, he spent his leisure time in the bars of the city and was a well-known client of a brothel run by Anna Sutherland. His first wife eventually divorced him in 1890, but he then proposed to Anna and they were soon married. Anna had amassed a fortune, but she was coarse in manner and language and therefore somewhat miscast as the wife of a respectable doctor—but in any case that marriage was not destined to last. Anna was taken ill and after only few days she died, on 22 April 1892. The doctor who was called to certify the cause of death put it down to a brain haemorrhage. Buchanan inherited her $50 000, equivalent to more than $1 million today.

Anna's former partner was upset to find she had left him nothing in her will and he told a reporter, Ike White of the *New York World*, that he suspected she had been murdered. Ike sensed a good story and went to interview Buchanan, only to discover that he had returned to Nova Scotia and had promptly remarried his first wife. Further enquires at Anna's brothel revealed that Buchanan had once boasted that he would be able to disguise a death by morphine, of which the most obvious symptom is pinpoint pupils. He had told how this could be done when he referred one evening to the trial of Carlyle Harris, who had murdered his wife in 1891. It was her pinpoint pupils which had alerted the doctors and detectives to her death as being due to an overdose of morphine, and so it proved. Harris was convicted of her murder and was executed in the electric chair.

Reporter Ike now began his own investigations into Buchanan and tracked down a former employee of Anna who remembered him commenting on the Harris case, and saying that if only he had had the sense to put some belladonna drops in his wife's eyes before she died he would have got away with her murder. White also interviewed the nurse who had looked after Anna during her illness—and she recalled Buchanan doing exactly that. The *New York World* ran Ike's sensational story and urged the New York coroner to exhume Anna's body for further tests, and this took place on 22 May. Morphine was discovered in the body, and it was estimated that she had been given a dose of at least five grains (300 mg). Buchanan was arrested and brought to trial in March 1893. It was during the trial that a cat was produced in court and injected with morphine and drops of atropine placed in its eyes. It was demonstrated to the jury that after it had died its pupils had not in fact contracted to pinpoint size. The defence counsel countered the scientific evidence with their own expert, but when Buchanan himself went into the witness box he destroyed his case with evasions and lies. He was found guilty, and executed in the electric chair on 2 July 1895.

3.10 The Attempted Murder of Alexandra Agutter

So strong is the bitter taste of atropine that it can be detected in concentrations as low as 100 ppm (0.001%). The would-be murderer Paul Agutter knew this, so he devised what he thought would be an excellent cover for its bitterness by adding it to tonic water, in which he assumed the taste of the quinine would mask it. As he was a biologist familiar with atropine he knew exactly how much constituted a fatal dose, and how much he would need to add to the other bottles of tonic water which he laid as a false trail, so that those drinking them would become ill but there would not be enough atropine to cause death. His story ranks alongside Crippen's as a fascinating account of a carefully planned but bungled murder.

At school Paul Agutter had shown himself to be highly intelligent and after leaving university he went on to pursue an academic career. In 1994 he was 49 years old and was married to another academic, Dr Alexandra Agutter, who taught English; they had two children. He lectured in biology at the School of Life Sciences of Napier University, located in Colinton Road in the southern part of Edinburgh. The Agutters lived at Kilduff Lodge in the historic village of Athelstaneford in East Lothian, about 20 miles east of Edinburgh. Athelstaneford is famous as the place where the Scots defeated the Angles in 823 AD, the battle being fought under a blue sky with clouds in the shape of a great white cross. It was this which led to the adoption of the saltire, the white St Andrew's Cross against a blue background which forms the Scottish national flag, and there is a commemorative stone to this effect in the church burial ground.

To outward appearances Agutter's life was one to be envied, but from his point of view things looked very different. Indeed in the early summer of 1994 Agutter had sent a message to Dr Ross Langlands, a general practitioner he knew, saying that he was so depressed that he was thinking of taking his own life. Langlands went to see Agutter at his home and was told that Agutter's marriage was finished, that he had financial problems, and that he was having an affair with a younger woman who wanted him to leave his wife and live with her. Such information passing between a patient and doctor is of course confidential and cannot be divulged unless both parties agree, but we know what was said that day because Agutter later signed a mandate allowing the doctor's notes to be released to the police. They were referred to at his trial as indicating a man with unresolved problems in his life, problems which he believed he could overcome by resorting to murder.

Agutter was fast approaching those dangerous years in middle age when a man's sexual desires and ambitions may again burn as strongly as they did in his youth. This testosterone-driven behaviour is sometimes referred to as the male menopause, and Agutter was certainly experiencing it. His new love was Carole Bonsall, a mature student at Napier University, and he wished to marry her. Divorce would have been the normal course of action but that would have left Agutter with no home and very little income, so he decided instead to remove his wife in another manner, and to achieve it using his knowledge of poisons.

Agutter never confessed to trying to murder his wife, so we can only surmise what his plans and *modus operandi* were in the summer of 1994. However, much can be deduced from the evidence put forward by the prosecution at his trial early the following year.

Agutter was well aware that the poison he chose would be discovered at autopsy, so the finger of suspicion had to point at someone else. Who better to blame than a nonexistent person? Agutter was an accomplished chess player, involving strategic thinking and planning several moves ahead, and his sophisticated plot to murder his wife clearly demonstrated these features. In actual fact Agutter was too clever by half.

Why he chose atropine as the poison with which to kill his wife was never revealed, but it was probably because he knew he could obtain it without suspicion. Within the University of Napier was a highly rated biomedicine research group whose speciality was toxicology, and Agutter was a member of it. Atropine is available from suppliers of research chemicals such as Sigma Aldrich, and it would have been easy for him to order it. In its 2005 catalogue appears atropine sulfate, costing £12.60 for 5 g. Also in the catalogue are atropine methyl bromide, slightly cheaper at £12.10 for this quantity, and atropine methyl nitrate, which is much more expensive at £25.80. These two variants are less toxic, which is probably why Agutter decided on atropine sulphate, as his poison of choice—not that he did anything so foolish as to place an order with Sigma Aldrich. He knew there was already a bottle of this chemical in the department from which he could take what he needed without it being missed or its possession being traceable to him. Indeed at his trial he admitted to having handled atropine as part of a biological experiment at the university three years previously, but claimed that he had not handled it since.

Agutter's plan was to put a fatal dose of atropine in his wife's gin and tonic one evening. To make her death seem the act of a random killer he also spiked several other one-litre bottles of tonic water with atropine and replaced them on the shelves of the Safeway supermarket at Hunter's Tryst in southern Edinburgh; this store was only a short drive from his office at Napier University. The date was Wednesday, 24 August 1994. A CCTV camera in the

Safeway supermarket captured Agutter in the store that day, although he was not seen actually putting the poisoned tonic water back on the shelf. The camera covering the drinks area was working, but unfortunately its video recorder was not. However, a part-time shelf-stacker at the store, who happened also to be a student at Napier although not taught by Agutter, later said that he had seen a man putting several bottles back on the shelves and was able to identify the man as Agutter.

Agutter kept one of the bottles of tonic water for himself and to it he added atropine at the fatal dose, and on Saturday evening, 28 August, he poured his wife a generous gin and tonic. Unfortunately for his plan, his wife found the drink too bitter and so drank only part of it. Even this might have been enough to kill her, and it was certainly enough for the symptoms of severe atropine poisoning to appear. When she tried to stand up she felt dizzy and collapsed on the floor. Within five minutes she was complaining of a pain in her throat, and was clearly in trouble and having hallucinations, which she later described as though the things around her were made of gossamer silk.

Agutter promised to ring for help but he merely called their local doctor, who was not at home, so Agutter left a message on his answering machine, making it sound urgent. So far so good, he must have thought; things were going to plan. However, unbeknown to him the message was picked up by the locum practitioner on duty that evening. He saw that help was required and went straight round to the Agutter's home. He also called for an ambulance. When the doctor and the ambulance arrived at the house it was obvious that Mrs Agutter was seriously ill, and they conjectured that she had been poisoned by something she had been drinking. Not only did the ambulance man take possession of the bottles of drink, he also poured the remains of Mrs Agutter's gin and tonic into an empty jar and took that away for analysis. Meanwhile Mrs Agutter was rushed to hospital, where her life was saved, although she remained seriously ill for some time.

By a strange coincidence one of Agutter's poisoned tonic water bottles ended up in the home of Dr Geoffrey Sharwood-Smith, a consultant anaesthetist who was familiar with the symptoms of atropine because he had employed the drug and seen its effects. His wife Elizabeth, and 18-year-old son Andrew, drank some of the poisoned tonic and they too were taken ill and rushed to hospital.

The anaesthetist correctly deduced that they were suffering from atropine poisoning and informed the hospital. Indeed, Sharwood-Smith wondered if the police might suspect him of trying to poison his family, because he had access to ampoules of atropine. (He actually discussed this possibility with Agutter when the two of them later met.)

In the last week of August 1994 eight local people were admitted to hospital suffering from atropine symptoms, including Mrs Agutter, the intended victim. The common factor linking these rare cases of atropine poisoning was that they had all consumed tonic water from the same store, and clearly all were victims of some criminal who it was assumed had done this in order to blackmail the supermarket. Such crimes are not unknown. Chief Superintendent John McGowan of the Lothian and Borders Police was put in charge of the investigation.

A press conference was called by the company, and citizens of Edinburgh were asked to return all bottles of tonic water they might have bought from the Hunter's Tryst Safeway. Those that were returned were tested and six of them contained atropine. The poisoned tonic water started a nationwide scare, and 50 000 bottles of Safeway's own brand tonic water were taken from the shelves and destroyed. Meanwhile the story became front page news and a dedicated helpline set up by the police was inundated by hundreds of callers. A 25-year-old man wrote to a local newspaper saying he had doctored the bottles of tonic. He was traced and arrested under the Mental Health Act and charged with wasting police time.

When the police investigated his wife's poisoning, Agutter even gave a press interview acting the part of an anxious husband and parent, saying his "wife and daughter could have been killed" and adding that the attempted murder of his wife was "outside anything I can understand. If anyone knows who the culprit is they should go to the police, and the sooner the better." Agutter contacted the Sharwood-Smiths, visited their home and even had a meal there, all the time maintaining the pretence of sharing their grief. He even had the temerity to write a letter in which he said, "Please accept my warmest sympathy and my assurance that the poisoner will give himself away some time, and that our system of justice very seldom convicts the innocent." Sharwood-Smith was unsettled by Agutter's visits, however, and after the trial he told a *Scotland on Sunday* reporter that he had wondered at the time whether Agutter might be the person responsible for the poisoned bottles.

What Agutter had failed to do was dispose of the evidence in his own home. Analysis of the tonic water in the bottle removed from there, and the remainder of Mrs Agutter's gin and tonic, showed that it contained considerably more atropine than the other bottles from the supermarket. Too late, Agutter realised that he should have thrown away the poisoned drinks and retained a second bottle of tonic water with less atropine in it to give to the investigators.

Agutter was eventually arrested and brought to trial, starting on 24 January 1995 before Judge Lord Morrison. On 21 February he was found guilty of attempted murder and jailed for 12 years, of which he served seven. The most telling evidence against him was that of the forensic scientist, Dr Howard Oakley. He had analysed the contaminated tonic water bottles and found there was much more atropine in the bottle of tonic used to make Mrs Agutter's drink. This contained almost 300 mg of atropine, so a third of this in a large gin and tonic would have provided a fatal dose. It was later deduced that Mrs Agutter had probably drunk about 150 ml, thereby receiving a dose of around 50 mg. Dr Oakley said that the level of atropine in the other bottles was significantly less, ranging from 11 to 74 mg per litre.

A police witness at the trial said that when he confronted Agutter with the evidence he had admitted buying two bottles of tonic water at the Hunter's Tryst store and said he noticed that the seal on one of the bottles was already broken. The police took Agutter's car away for analysis and traces of atropine were found on a cassette case. His explanation was that the bottle must have leaked and had transferred the atropine from the bottle to the cassette.

After finding him guilty Judge Morrison said: "This was an evil and cunningly devised crime which was not only designed to bring about the death of your wife, but also to cause great alarm, danger, and injury to the public." He sentenced Agutter to 12 years in jail. Agutter went on to appeal on the grounds that the evidence was entirely circumstantial and that the forensic samples had been mishandled by the police. The three Judges of Appeal were not swayed, and concluded that the verdict had been sound and secure.

While in Glenochil prison, near Stirling, Agutter became a reformed character and began to devote himself to teaching the other inmates to read. He worked in the prison library and made a special effort to help his fellow prisoners gain basic skills. He told the media he would continue to do this after his release. One prisoner he helped spoke of him as a very plausible and polite man. Agutter even submitted an essay about his work as his entry for a competition run by the Prison Reform Trust, part of which read as follows:

> "Virtutally illiterate men have learned to read and write to at least early primary school standard . . . and several have begun to read with more enjoyment and to write letters to family and friends with great facility. The personal satisfaction is worth emphasising. I have become passionately committed to Adult Basic Education tuition. I am so convinced of its value that I hope in the future to be in a position to foster its development in prisons throughout the UK."

A policy officer of the Trust, Joe Levenson, commented on the importance of Agutter's work and pointed out that two-thirds of prisoners had such a low level of literacy and numeracy that it was difficult for them to find employment. Agutter was in the process of redeeming himself. Meanwhile, in 1998 Alexandra Agutter was granted a divorce.

When Agutter knew he was about to be released on licence in September 2002, he phoned his former wife to say that he was innocent and he expected to come back and live with her in Athelstaneford. She naturally wanted nothing more to do with him and was granted an interdict to prevent him contacting her. Consequently, on being paroled Agutter left Scotland and went to live with his ageing parents in Derbyshire in England. He began to apply for work and eventually was offered part-time employment at the University of Manchester, teaching evening classes. And what subjects did he teach? They were philosophy and medical ethics. And why not?

CHAPTER 4

Diamorphine and the Dr Jekyll of Hyde

A word shown in **bold** *indicates that further information will be found in the Glossary*

4.1 Dr Shipman's Addiction

Joan May Melia had no family and lived alone in social housing in Hyde, near Manchester. She was 73 and suffered from emphysema, a lung condition which makes it difficult to breathe, and whenever she had a chest infection she saw her doctor—as she did on 12 June 1998. That afternoon he called at her home to give her what she assumed to be an antibiotic injection, but curiously it sent her into a deep sleep and within a few minutes she had stopped breathing. At 5 pm Joan was discovered by a friend, who realised she was dead. He rang her doctor, who arrived soon after and immediately diagnosed death from pneumonia and entered this on the death certificate. It should really have said 'poisoned with diamorphine', because the doctor was Harold Shipman, Britain's most prolific serial killer. Joan Melia may have been his 200th victim—or was she his 300th, or even his 400th? He had stopped counting several years ago.

Dr Harold Frederick Shipman—Fred to his friends—had been a much loved and well respected family doctor for 25 years. The fact that he was a drug addict, serial killer, thief and forger escaped virtually everyone's notice. He had been cured of his addiction to drugs,

Molecules of Murder: Criminal Molecules and Classic Cases
By John Emsley
© John Emsley, 2008

but not of his addiction to murder, which went undiagnosed despite his needing a fix at least once a month. Even after this had been suspected and reported to the police by a fellow doctor, these suspicions were not properly followed up. In the end it was Shipman's amateurish performance as a forger which led to his downfall.

4.2   Morphine and Diamorphine

"Here was a panacea for all human woes; here was the secret of happiness about which philosophers had disputed for so many ages; happiness might now be bought for a penny; portable ecstasies might be had corked up in a bottle and peace of mind could be sent in gallons . . ."

So wrote the Edinburgh author Thomas de Quincey in his book, *Confessions of an English Opium Eater*, written in 1821. He was extolling the effects of laudanum, which was the name given to a solution of opium in alcohol, of which the active ingredient was **morphine**. The poet Samuel Taylor Coleridge was in the habit of consuming a pint of laudanum every week, and maybe he had done so when in the 1790s he wrote his most famous works, the awe-inspiring *The Rime of the Ancient Mariner* and the mysteriously exotic *Kubla Khan*. When morphine permeates the brain then there is no pain, no worry and no sadness. Life takes on a rosy glow and the user experiences that rare human emotion we describe as happiness—albeit for only a few hours. What de Quincey didn't mention, but Coleridge discovered, was that drinking laudanum was rather like smoking tobacco: once you were hooked it wasn't at all easy to stop. Nevertheless, doctors prescribed laudanum as a painkiller and a cure for diarrhoea, because it calmed the muscles of the intestines. A non-medical way of enjoying opium's effects was to inhale the vapour, in other words to smoke it in the manner of the fictional detective Sherlock Holmes. Addicted smokers frequented the notorious opium dens located in the back streets of many cities, especially in China and the Far East.

Opium comes from the opium poppy, *Papaver somniferum*, and its name is taken from the Greek word *opion*, meaning poppy juice. It is an annual plant which grows to over a metre in height and it appears to have originated in Asia Minor (modern Turkey) where it was already cultivated for its opium 5000 years ago. Some strains of *Papaver somniferum* yield more opium than others, although it is important that the juice is extracted at the right time, as soon as the greenish seed pods form and they are still unripe. The juice is obtained by cutting into the pod capsule, when it oozes out and coagulates to form a pale brown gum which after it is dry can be ground to a powder. Opium contains up to 20% morphine and 2% **codeine**.

Opium was first investigated chemically in 1804 by Armand Séquin, who extracted from it a crystalline material which he named **morphine**, after *Morpheus*, the Greek god of sleep. Around the same time, the German pharmacist Friedrich Sertürner also isolated it and correctly linked it to opium's narcotic power. In fact, morphine is ten times more powerful than raw opium. It became particularly important as a medical treatment once the hypodermic needle had been invented in 1853, and it then began to find use as the strongest painkiller available, proving invaluable as a battlefield medicine in the Crimean War (1853–56) and the American Civil War (1861–65).

The juice of the poppy is still collected in the traditional manner to make the heroin illegally available on the streets. The pharmaceutical industry, which produces this drug legally as diamorphine, harvests the poppy in a different manner using crops grown in places such as Tasmania, whose farmers supply about a quarter of the world market for the legal product. The poppies are mechanically harvested in two stages. First the seed pods are removed and their contents can be used, for example, in poppy seed cakes and bagels. The straw is then cut and the 0.5% morphine it contains is extracted using solvents. The seeds also contain traces of morphine, sufficient in fact for a person who eats them to test positive for the drug the following day, as some people have found to their cost in the USA, where testing is mandatory for government employees and in jobs such as nursing or those involving driving.

Opium is mentioned in ancient Egyptian papyri. An ivory 'pipe' dating from 1200 BC has been unearthed in Cyprus and is thought to have been used for inhaling opium. Dioscorides, a Greek physician who practised during the first century AD, gives an account of the way opium was produced and used. Cultivation of the opium poppy spread throughout the Middle East, and eventually to India and China. In the 1500s it was popularised in Europe by Paracelsus, who showed that a solution of opium in alcohol was an effective painkiller, and he called the tincture laudanum, from the Latin *laudare*, to praise. Meanwhile opium became particularly popular in China following the banning of tobacco in 1644, and like tobacco it was smoked and inhaled.

The British carried on a major trade in opium between India and China. When the Chinese government tried to stop this by seizing and destroying stocks of the drug it led to the two Opium Wars (1839–42 and 1858–60), the first of which ended with a peace conference which legalised opium and ceded Hong Kong to the British. The second war was sparked by the sinking of a British ship by the Chinese.

Opium always caused harm to the societies where it was readily available and was used recreationally, but it was only in 1914 that an international conference agreed to curb its production world-wide. However, there was no way in which the poor farmers in countries such as Pakistan, Afghanistan, Turkey, Burma and Thailand were likely to stop growing such a valuable cash crop, and they have continued to do so to the present day.

Morphine can be taken into the body in a variety of ways: by smoking, by sniffing it as a powder, by drinking it as a solution, or by subcutaneous, intramuscular or intravenous **injection**. It acts directly on the central nervous system, with a half-life in the body of about two or three hours. Morphine is metabolised by the liver and excreted in the urine. It is highly addictive when used for any length of time, and the main withdrawal symptoms include back pain, hot flushes with alternating feelings of cold (hence the expression, 'cold turkey'), sweating, and kicking spasms in the legs (hence the phrase, 'to kick the habit'). In the UK morphine is a Class A drug with regard to its illegal possession, and in the USA it is a Schedule II drug, meaning that it can be used under supervision although it is known to be addictive.

The medical use of morphine varies from country to country. In 1996 it was estimated that in Denmark around 3 people per 1000 were given morphine to control pain, whereas in the UK the figure was lower at around 2 per 1000. In some countries it is hardly used at all, out of fear that it will lead to addiction. Provided it is pre-scribed briefly to reduce pain, the risk of addiction is actually quite low; according to a 1980 paper in *The New England Journal of Medicine*, only 1 person in 3000 becomes addicted, and the risk of addiction is even smaller if the patient can control his or her dosage using a drip.

Morphine targets the opioid receptors, of which there are three types, known as δ, κ and μ. These were discovered in 1973 in rat brains after labelling opium-type drugs with tritium. In 1996 it was shown that morphine specifically targets the μ-opioid receptors. Once these receptors are out of action even the most excruciating pain fails to register, and in addition there is a wonderful sensation of euphoria. (The brain produces its own chemicals as a way of controlling continuous pain. These are short-chained amino acids known either as enkephalins, which target the δ-receptors, or endorphins, which mainly target the κ-opioid receptors.) When mice were specially bred without μ-opioid receptors they were found to be totally unresponsive to morphine.[i] Rather unexpectedly, in 2004 researchers at the Martin Luther University and the Leibniz Institute of Plant Biochemistry in Halle, Germany, reported that human cells were able to generate morphine, which they demonstrated by using the precursor tyramine with inbuilt carbon–13 isotopes, and oxygen gas consisting only of oxygen–18. These were seen to be present in the morphine which was later detected, the implication being that the human cells were synthesising this molecule. It is unclear why they do this.

The safe dose for morphine is 20 mg. An accidental death with morphine occurred in Wollongong, New South Wales, in 2006. A doctor provided a patient suffering from severe back pain with 120 mg ampoules of morphine tartrate in place of the more usual morphine sulfate, which comes in 20 mg doses. The patient injected the entire contents of one ampoule into his thigh and died 12 hours later. The doctor admitted culpability, was found guilty and given a suspended prison sentence. An alternative to injection is the oral form of morphine sulfate, available as the medication Kapanol. This consists of a capsule containing tiny coated pellets of the drug, which is slowly released over 12 hours. The introduction of these controlled-release tablets in the 1990s increased world demand for morphine several-fold.

When morphine targets the μ-opioid receptors in the brain it slows respiration. This increases the danger from an overdose since

---

[i] The mice were bred using embryonic stem-cell technology from which the relevant gene had been removed.

it interferes with the mechanism used by the brain to adjust the rate of breathing according to the amount of carbon dioxide dissolved in the blood. This gas is a by-product of the body's use of its most important energy source, glucose, and the more we use the faster we need to breathe. If the brain's ability to detect carbon dioxide is suppressed, the rate of breathing may slow to a level which is unable to sustain life.

4.3   Chemically Modified Morphine

Morphine is a complex molecule which can be modified in a number of ways, each change having a profound effect on its action. Non-addictive painkillers such as **codeine** and **hydromorphone** are produced from morphine, as is the cough-suppressant **pholcodine**. So also is **diamorphine**, better known as heroin, but this is as addictive as morphine itself.

During the last century pharmaceutical companies synthesised hundreds of morphine derivatives in the hope of finding one which had the pain-killing power of morphine but without its addictive properties. Unfortunately, in many derivatives there seems to be a trade-off between the two. Codeine (methylmorphine sulfate) is one such example; although it has no addictive properties it is much less effective as a painkiller since it does not bind so efficiently to the μ-opioid receptors. In the body it is slowly converted to morphine but not at a rate which allows it to build up to significant levels, which is why the safe dose for codeine is higher, at 60 mg. Codeine accounts for 70% of the legally grown morphine.

Pholcodine (morpholinoethyl morphine) is even safer, but is not an effective painkiller. It suppresses the muscles causing a tickling but unproductive cough, and is an ingredient in some over the counter cough mixtures.

Diamorphine (diacetylmorphine) is both powerful and addictive. It was first produced in 1874 and marketed as a cough suppressant. Because it was so potent it was given the name heroin, from the Greek word *hero*, signifying a human who had won the approval of the gods on account of his courage. Whatever morphine could do, heroin could do better. By adding two acetyl groups to the molecule it becomes more soluble in lipid tissue and crosses the **blood–brain barrier** more rapidly. The acetyl groups are removed by enzymes in the blood and the brain, and the resulting morphine molecule can then get to work. Diamorphine is highly addictive and is now only prescribed to patients in the final stages of terminal illnesses, for whom addiction is not really an issue. Diamorphine brings them relief and a dignified end to life.

In the USA another morphine variant, hydromorphone, is preferred. This is a keto-derivative of morphine which is even stronger than diamorphine as a painkiller and apparently much less addictive. A mere 2 mg of hydromorphone is as effective as 10 mg of diamorphine or 20 mg of morphine.

While small doses of these modified morphine drugs are not life-threatening, larger doses certainly can be. The normal dose of diamorphine is 5 mg given every four hours by subcutaneous or intramuscular **injection**. It is injected slowly into a vein on the back of the hand or in the crook of the elbow. A dose of 30 mg injected all at once into a vein will quickly kill an adult, or even less if they are frail or elderly. Some doctors were at one time known to give a dying patient a double dose of morphine or diamorphine to hasten death; some regarded this as euthanasia, others as assisted suicide, but many relatives saw it as a merciful release of a loved one from further suffering. Dr Hazel Biggs is the author of *Euthanasia: Death with Dignity and the Law,* and she concludes that annually around 18 000 people in the UK are helped in this manner by their GP, with considerable public support. An article in the *British Medical Journal* in 2004 spoke of a Dutch doctor charged with using diamorphine to murder a 77-year-old patient who had suffered a massive stroke. The doctor was acquitted, and indeed the Dutch Minister of Health admitted that so-called 'terminal sedation' was involved in about 10 000 cases a year in that country.

Naloxone is morphine with the methyl on the nitrogen atom replaced by an allyl propenyl group. This molecule can easily displace morphine from the μ-opioid receptor it is docked on to, yet naloxone does not have the ability to trigger the receptor in the same fashion. When life is in danger following an overdose of morphine or heroin, a patient will revive within minutes if given an injection of naloxone. Overdosing is most likely to happen with illegal heroin if this has not been sufficiently diluted. Heroin bought by drug users has normally been greatly diluted ('cut') with an innocuous powder such as lactose or dried milk and its heroin content is often below 5%. This is in contrast with uncut heroin, which may be 90% pure. Too much heroin and the user's life becomes at risk. Indeed, when I was a lecturer at King's College, London in the 1980s, one of the chemistry students and his girlfriend died as a result of injecting uncut heroin.

4.4 Dr Harold Shipman

In-depth psychoanalysis might have revealed how Shipman could be both a serial killer and a well-loved GP with a number of close friends, but he was always unwilling to take part in such an investigation. Nevertheless, there were two factors in his early years that might have influenced him. The first was the death of his mother from cancer and the relief she obtained from morphine injections. The second was the sensational trial of a doctor who murdered using morphine, but was acquitted—see the box below.

When Harold Shipman was 17 his mother, Vera, was diagnosed with lung cancer, and she died on Friday, 21 June 1963. Her suffering was alleviated by daily injections of morphine, which in those days had to be administered in the presence of a third party, and that was sometimes her son. He was impressed by the relief the injections gave his mother but when it was clear that her end was near he may have noticed her doctor giving her a double dose, and she passed away peacefully.

Dr John Bodkin Adams

When Shipman was a boy the newspapers had been full of the trial of the morphine murderer, Dr John Bodkin Adams, a GP attending the rich widows of Eastbourne. This is a seaside town famed for its mild climate, and therefore an attraction for retired wealthy folk. Scores of his patients left him money and other gifts in their wills, and he became a wealthy man. But there were others who suspected that some of his benefactors might have been given a little help on their way to the grave. In 1957 Bodkin Adams found himself in court charged with the death of a former patient, Edith Morrell, who had died in 1949 and left him her Rolls Royce and a chest of valuable Georgian silver plate and cutlery. (In the week of her death Bodkin

Adams had written prescriptions totalling 2600 mg morphine and 2500 mg diamorphine.) Because her death had occurred so long ago it was not difficult for Adams's defence counsel to show that the nurses who were on duty at the time the doctor gave the injections could no longer be regarded as reliable witnesses. The defence demonstrated that the nurses' evidence in court was in conflict with the notebooks they had kept at the time. Bodkin Adams was acquitted, although later that year he was convicted of forging prescriptions, making false statements on cremation forms, and misusing dangerous drugs. For this he was fined £2400, equivalent to as much as £50 000 today. It now seems almost certain that over a 30-year period he killed 160 of his patients, inheriting from some of them quite remarkable sums of money. In 1935 he had a bequest of £7000, equivalent to £1 million today. John Bodkin Adams died on 4 July 1983, leaving an estate worth £400 000, perhaps £1.4 million today. He murdered for financial gain, whereas Harold Shipman, although more prolific as a killer, murdered only once for money—and that murder was his downfall.

Shipman was born on 14 January 1946, the second child of Vera and Harold Shipman; he had an older sister. The family lived at 163 Longmead Drive, Nottingham. Fred, as he was known, was bright enough to be selected to attend High Pavement Grammar School, where he excelled at sport, especially rugby. He was successful enough academically to stay on for two extra years with a view to studying medicine at university. His first attempt to gain admission to Leeds University was unsuccessful and he remained at school for an additional year, to be accepted by the University Medical School in 1964.

In his second year, in May 1966, Shipman failed to use a condom when making love to his girlfriend, Primrose Oxtoby. She was an attractive, bubbly 17-year-old who worked as a window dresser in a store in the city; Fred was the archetypal handsome, sporty medical student. When Primrose found she was pregnant the two were quietly married at Barkston Ash Register Office on 5 November 1966, and baby Sarah was born three months later, on 14 February 1967. (Shipman's later children were Christopher, 1970, David, 1979, and Sam, 1982.)

In 1970 Shipman graduated and was provisionally registered by the General Medical Council. Before being fully registered as a doctor he had to work for a year in a hospital, including six months involving surgery, and he did this at Pontefract General Infirmary, southeast of Leeds. Shipman remained at the hospital for a further two and half years and qualified as a senior houseman, with diplomas in child health, obstetrics and gynaecology. While he was at Pontefract it is just possible that he committed his first murder. A health worker later remarked that at least one of his elderly female patients had died unexpectedly, but we cannot know for certain that he was responsible.

Shipman decided to move into general practice and in March 1974, aged 28, he obtained a temporary position at Todmorden in the hilly Pennine area of West Yorkshire. He joined a new-style group practice in a purpose-built facility. Shipman threw himself into his new role—nothing seemed too much for him—and within a few months he was offered a permanent position. He and Primrose bought a 1930s semi-detached house in the Sunnybank area of the town. Shipman became an active member of the local community, to the extent of joining the society that was in course of restoring the Rochdale Canal. What no one realised was that he had become addicted to pethidine, which he obtained by prescribing more of the drug than was required and retaining the balance for his own use. It was not difficult to acquire pethidine, since it was widely prescribed to ease labour pains.

His frequent prescriptions for pethidine attracted the attention of the government official responsible for monitoring drug use, however, and in June 1975 the official asked the Police Drug Squad to check with Todmorden pharmacists to confirm that Shipman was indeed prescribing a large amount. When Shipman was questioned he confessed to having taken pethidine at a student party, but insisted that this was merely a one-off event and had not been repeated. Such were the glowing reports of the unstinting effort he was applying to his job, that it was concluded there was little to suggest that he had behaved in any way irresponsibly. Sometimes after taking the drug he passed out, even in his surgery, but this was excused as being a form of epilepsy.

However, time was running out for Shipman's abuse of pethidine, and when one of the receptionists in the group practice was shown the dangerous drugs book at her local pharmacist and witnessed the enormous quantities of pethidine that Shipman had prescribed, she mentioned the fact to another doctor in the practice. It was obvious that he needed it for his own use. Eventually he admitted his addiction to his colleagues, and when they refused to hide what he had been doing he at once resigned, had himself taken off the register of GPs, and was admitted to a private clinic for treatment. It was noted there that the veins in his thighs where he was injecting the drug were in a collapsed condition, suggesting that he had been injecting himself for at least five years. He had been actually been using as much as 600 mg of pethidine daily.

By the end of 1975 Shipman was able to leave hospital and was regarded as cured. Two months later he appeared before magistrates and pleaded guilty to three offences of obtaining pethidine for his personal use, and of forging prescriptions to obtain it. He asked for 74 similar offences to be taken into account, and it became clear that during his time in Todmorden he had stolen a total of 70 000 mg of pethidine. He was fined £600, equivalent to about £5000 today. He then moved to a desk job in the Newton Aycliffe Health Centre in Durham, without access to listed drugs, and he appears to have been cured of his addiction—indeed no subsequent investigation has suggested otherwise.

How many of his patients did Shipman kill while at Todmorden? During the 16 months he worked there he certified 22 deaths. Shipman was actually present, or had just left them, when nine of his patients had suddenly died. Indeed, on one day alone, 21 January 1975, he appears to have murdered Elizabeth Pearce (84) at 4 pm, Robert Lingard (62) at 7:30 pm, and Lily Cossley (73) soon afterwards. To the last of these he gave an injection 'to calm her down', or so he mentioned to a relative. The enquiry held after Shipman was later convicted of murder concluded that only one death in Todmorden could with certainty be attributed to him, that of Eva Lyons, whom he murdered on 17 March 1975. She died while her husband was present and Shipman gave her diamorphine through her drip. Although the enquiry concluded that the other patient deaths at Todmorden could not be conclusively attributed to Shipman, some almost certainly were.

Life at Newton Aycliffe was not what Shipman wanted and he looked around for a different GP position. In 1977 he applied to the Donneybrook Medical Practice in Hyde.[ii] He was interviewed and offered the post, and began work on 1 October 1977. He so impressed the other doctors in the group that after a six-month trial period his position was made permanent. He quickly gained a reputation for hard work, and was willing to visit patients at any time. Not only did he write the prescriptions for patients, he was quite prepared to collect the drugs himself from the nearby pharmacy and deliver them personally. Sometimes he even turned up unannounced, ostensibly to check how his patients were responding to treatment. Shipman's old style doctoring was much appreciated by the elderly and infirm, who sang his praises and saw him as an asset to the community. The Shipmans lived in a modest house, 15 Roe Cross Green, in the pleasant town of Mottram a few miles east of Hyde. Outside surgery hours Shipman played an active part, as a school governor, was a member of the Parent Teachers'Association at Longdendale High School which his children attended, and was even its chairman for a while. He also ran the rugby club and sat on various medical committees.

In August 1978, a few months after the end of his probationary period, Shipman called to see 86-year-old Sarah Marsland. While he was giving her an injection of diamorphine her daughter arrived. (The diamorphine was left over from that prescribed for a cancer patient who had died the month previously.) The daughter was told that her mother had just had a heart attack and that Shipman was trying to revive her, but he was apparently unsuccessful, because she died within a few minutes. (Shipman murdered the daughter 20 years later, in 1998.) Later that month Shipman murdered Mary Jordan, and in December that year, Harold Bramwell. His career as a serial killer was now back on track and continued for another 20 years, slowly at first but with increasing speed as the years went by.

Shipman's typical victim was an elderly woman living alone, and in all cases the killings seemed motiveless. The victims were not in any way people in great pain at the end of their lives and who

---

[ii] Hyde became infamous in the UK in the 1960s as the place where Ian Brady and Myra Hindley lived, and where they abducted, tortured and murdered five young children before burying them on the moors. Some of their graves have still to be located.

wished to die. Nor did he derive financial benefit from their deaths, although he did sometimes help himself to items of jewellery. He killed them using either morphine or diamorphine.

During the two years, 1981 and 1982, almost all of Shipman's patients who died did so of natural causes, totalling 14 in 1981 and 12 in 1982. In 1983, 11 died—and two of these are now known to have been killed deliberately. From that time onwards he averaged roughly one murder a month, although not with any regularity; sometimes he might kill three within a week, as he did around Christmas 1984. He now had a *modus operandi*, and one which would provide him with an alibi. If the victim had a gas fire in their living room, this would be turned full on, with the body sitting in a chair. The rate of cooling of a body can indicate how long a person has been dead, but this is clearly affected by the temperature of the surroundings. When called out the following day to one of his victims, he could safely measure the body temperature and write an estimated time of death on the certificate which was some time later than his visit the previous day.

During 1985 eight of Shipman's patients died natural deaths and 11 were murdered. In June that year he even killed two on the same day, and again there was a spate of killings around Christmas. During 1986 and 1987 the number of killings was eight each year, and again in 1987 there was a cluster of three in the Christmas period. In 1988 four of his patients died from natural causes and 11 were murdered, four within one week in February, and then none until September. In 1989, three patients died natural deaths and 12 were murdered, including one, Mary Hamer, whom he killed in his surgery. He injected her in the examination room and while waiting for her to die he saw other patients. He then 'discovered' that Mary appeared to have suffered a heart attack.

It was in 1989 that Shipman came close to being uncovered, when in November he murdered 85-year-old Joseph Wilcockson. Hardly had he left the body when a district nurse arrived to change the dressing on an ulcer on Mr Wilcockson's leg. She let herself into his home to find him apparently asleep in his chair. His body was still warm, so she rang Shipman, who came to certify that his patient was dead—but he did not mention that he had visited him only a short time previously. That was a close call, and Shipman did not commit another murder for 10 months. In 1990 he murdered only two people, and just one in 1991, but things were about to change, which would make his excessive prescribing of diamorphine obvious to his colleagues.

In 1992 the Donneybrook doctors decided to become a group practice, which meant that any of the doctors was equally likely to treat any of the patients. This would of course have opened up Shipman's medical notes to scrutiny, so he decided he would move out and start his own practice a short distance away. The local newspaper, *The Tameside Advertiser,* reported the opening ceremony in June 1992, when the ribbon across the door was cut by one of his youngest and oldest patients, a three-week old baby girl, Emma Myllot, and a 101-year-old lady, Minnie Ward. Some of the support staff moved with him, as did no fewer than 2100 of his patients—indeed he soon had a total of 3100 on his books and a waiting list wanting to join.

The murders could now resume, and they did so with increased frequency. In the years between 1993 and 1998 Shipman killed 140 of his patients, averaging two a month, although in 1992 he had murdered just one, Monica Sparks (72). In 1993 he made up for lost time by murdering no fewer than 18, each time injecting a 30 mg ampoule of diamorphine. On rare occasions a murder was actually witnessed, as on 22 July 1993 when he killed José Richards (74). Shipman was still drinking the cup of tea she had made for him when a friend called to visit her. Shipman admitted that he had just given her an injection, saying she had been in terrible pain after a heart attack, but that she had nevertheless died. He had an even closer shave a year later, on 17 May 1994, when he was disturbed while actually giving Mary Smith an injection. He stopped the injection, and all Mary suffered was a sleep lasting more than 18 hours from which she recovered with no lasting harm. He finally murdered her two years later. That near miss made him more careful, and he postponed his killing programme until Christmas that year, when he dispatched three patients in quick succession.

During 1994 he killed 12 patients, again murdering one at his surgery. In 1995 he killed 30, sometimes two in one day, and he altered his technique slightly. By injecting intramuscularly he knew that a victim would take longer to die, which would give him opportunity to get well away from the scene and to create a watertight alibi. On some occasions this gave the victim time to make a phone call, and some did, saying that the doctor had given them an injection of antibiotics to cure whatever was afflicting them. The following year, 1996, he murdered a further 30. In September, however, he unknowingly set a trap for himself. He had installed new software on his computer, and although he did not realise it, this recorded the date and time of each entry. This meant that when he returned from a killing and faked a victim's medical notes, making it appear they had had a heart condition for two or three years, he was actually providing proof of what he was up to.

Shipman's frenzy of killing reached its peak in 1997 with a tally of 37, beginning on 2 January, when he was called to a nursing home to see a patient with a high temperature. He diagnosed heart failure, and actually injected her with diamorphine while the deputy manager of the home looked on, saying that he was trying to kick start her heart with epinephrine. Later, Shipman was almost caught in the act when he murdered Betty 'Lizzie' Adams on 28 February, the day after she returned from a holiday in Malta, where she had been dancing every day, this being the love of her life. She went to see Shipman on her return because she had a cold and a cough. The following day he went round to her flat around 2 pm and gave her the fatal injection. While he was waiting for her to die, Lizzie's regular dance partner, Bill Catlow, arrived. Shipman said he was waiting for an ambulance to take Lizzie to hospital. When Bill tried to arouse her and couldn't, he pointed this out to Shipman, who pronounced her dead and then pretended to cancel the ambulance. Lizzie's cause of death was given as bronchopneumonia.

A few years previously Shipman had had another close shave when he went to see 81-year-old Maria West. She had rung the doctor, asking him to call to see her because the painkiller he had prescribed for her arthritis did not seem to agree with her. He promised her something stronger, and saw her during the afternoon of 6 March 1995. Although he did not realise it, Maria was being visited by a friend, who had discreetly gone into the bathroom when the doctor's car drew up outside. A few minutes later Shipman was confronted by the visitor, whom he informed that her friend had died just a minute ago—indeed there she was, sitting in her favourite chair as though asleep. What puzzled the friend was that the doctor made no attempt to resuscitate Maria.

Shipman's killings continued into 1998, when a further 18 patients were murdered. However, he was becoming careless and events were now beginning to overtake him.

4.5   Suspicions Aroused

Three people began to have suspicions that Shipman might be killing his patients: a local taxi driver, John Shaw, who specialised in ferrying old ladies around Hyde; a local funeral director, Alan Massey, who was concerned that so many of the people he was dealing with were Shipman's patients; and a local doctor, Dr Linda Reynolds, who was so worried about what Shipman might be up to that she informed the authorities.

The first of John Shaw's regular customers to die was Monica Sparkes, in 1992. He knew about her doctor because she spoke so highly of him—as indeed did many of his regulars, who talked about Shipman and said how lucky they were to have him as their GP. Then they would unexpectedly die. Shaw began to compile a list, and it became longer and longer. However, Shaw was not in a position to discuss his suspicions with anyone, even though there were eventually as many as 20 names on his list.

The undertaker, Alan Massey, was intrigued that at many of the homes he was called to the story was often the same: the person had died unexpectedly and was found sitting in an armchair. Massey commented to Shipman about the large number of his patients that he had either buried or cremated, but the doctor allayed his fears by being completely open, and even invited Massey to look through his medical files. The undertaker concluded that there was nothing to worry about. Massey's partner, Debbie, took their worries more seriously, however, and she voiced their suspicions when on one occasion she took a cremation form to the Donneybrook surgery for a second signature. (Two doctors must examine a body which is to be cremated and sign a form to confirm that the death is above suspicion.) Dr Linda Reynolds, a newcomer to the Donneybrook group, was intrigued, and quietly took a look at the death rate of Shipman's patients. She noticed that the group practice had more than three times as many patients as Shipman himself had, and yet the number of deaths he recorded was three times greater than the total for the practice. She calculated that a person on Shipman's list had a ten-fold higher chance of dying. Dr Reynolds called the local Coroner's Office on Tuesday, 24 March, and suggested that someone should investigate. The coroner, John Pollard, contacted the police and asked them to make discreet enquiries, without alerting Shipman.

A detective inspector was assigned to the case and interviewed Dr Reynolds, who informed him that two of Shipman's patients, Ada Warburton and Lily Higgins, currently awaited cremation and perhaps the coroner should order autopsies. The detective inspector was not really convinced there was anything to investigate, however, and the bodies were duly cremated. Dr Reynolds was vindicated later when eventually Shipman was brought to trial and convicted. Sadly she died of cancer soon after.

Shipman may have sensed that suspicions were being aroused and he paused for a few weeks. When he started again he seems to have targeted women who he knew were unlikely to be cremated, for example Roman Catholics, in which case a second doctor's signature was not required. This change in tactics was eventually to provide the most damning forensic evidence at his trial. Morphine and diamorphine only disappear as bodies decompose, and its presence in his victims' remains was a crucial factor in convicting him.

4.6   The Murder of Kathleen Grundy

In 1998 Shipman was 51 years old. He began to think in terms of early retirement and the possibility of moving to the south of France, where he could enjoy warm weather, good food and cheap wine—and no awkward patients to worry about. All he needed was enough money. So he decided to target Kathleen Grundy, a wealthy 80-year-old widow—her husband had been the Mayor of Hyde and had died 30 years previously. She not only owned a picturesque house, Loughrigg Cottage, on the hills overlooking Hyde, she also had other properties and investments. She could be worth half a million pounds. All Shipman needed to do was to forge her will, making him sole beneficiary, and then kill her—and that is precisely what he did.

Although she was an old lady, Kathleen Grundy was sprightly and active, and yet she could be rather a difficult person to deal with.[iii] She bought the food every week for an old folk's luncheon club. When its members turned up for their meal on Wednesday 24 June 1998 they were disconcerted to find that Kathleen was not there, and when they rang her home there was no reply. Eventually

---

[iii] Or so I was informed at the Manchester Literary and Philosophical Society by someone who had known her personally.

two of the club members went to Loughrigg Cottage and discovered Kathleen lying on her settee, apparently asleep. She was however already cold to the touch, and must have died earlier that morning. They rang Shipman, who they knew was her doctor, and stayed until he arrived. They were a little surprised that he hardly examined Kathleen before he announced that she had died of a heart attack, but he wrote 'old age' as the cause of death on the death certificate.

Kathleen had been flattered when Shipman said he would like her to take part in research carried out by a colleague at Manchester University into why some people in old age were so much healthier than others. She would need to sign a form that she was willing to take part in the survey, and her signature had to be witnessed by two other people. It happened that Paul Spencer and Claire Hutchinson were waiting to see the doctor, and they were asked to sign that they had witnessed her signature and added their addresses. The date was 10 June.

Shipman now forged Kathleen's will and traced the three signatures on to it. On a standard Last Will and Testament form, and using an old typewriter, some of whose keys were misaligned, he typed in capital letters:

ALL MY ESTATE, MONEY AND HOUSE TO MY DOCTOR. MY F MILY ARE NOT IN NEED AND I WANT TO RE ARD HIM FOR ALL THE CARE HE HAS GIVEN TO ME AND THE PEOPLE OF HYDE. HE IS SENSIBLE ENOUGH TO HANDLE ANY PROBLEMS THIS MAY GIVE HIM. MY DOCTOR IS DR.H. SHIPMAN 21MA K T ST HYDE CHESHIR SLI4 2AF

Maybe Shipman supposed that an old lady like Kathleen would make such a clumsy job of typing, and hence the poor punctuation, missing letters and incorrect spacing. In the Residuary Gifts section of the will it indicated that if she were to outlive the doctor then her estate should pass to her daughter, Angela. Shipman ticked 'cremation' as the method of disposing of the body. The will, dated 9 June 1998, was eventually sent off to the solicitors, Hamilton Ward & Co, who were appointed her executors on Wednesday, 24 June, the very day that Shipman had made an early morning call on Kathleen and injected her with the fatal dose of morphine.

When her mother died, daughter Angela did not suspect that there was anything wrong with the cause of death that Shipman had written on the death certificate, which was simply 'old age'. Angela grieved for her mother, and arranged her funeral and her interment in the cemetery at Hyde on 1 August 1998. She was not cremated, because the new will sent to the solicitors had been filed away unread.

What Shipman had not counted on was the indignation which Kathleen's daughter, who was herself a lawyer, felt on discovering that she and her children were not to inherit her mother's estate. Angela knew how much her mother was worth and she had in her possession a will which her mother had written some years previously, leaving the bulk of her estate to family members. It therefore came as a shock when she received a phone call from the solicitor in Hyde to inform her that he was in possession of a new will, which had reached him the day her mother died. He had been surprised to receive it, because her mother was not one of his clients. The will had been accompanied by a brief note signed by Kathleen Grundy and both letter and will had been filed away. A week later another letter arrived, this time from a Mr Smith to inform the solicitor that Kathleen Grundy had died and saying that he was the person who had typed out Kathleen's will for her. Mr Smith was of course Shipman. Only days after Kathleen Grundy's death, Shipman told another doctor in Hyde that he was about to come into a substantial inheritance, which in the near future would enable him to retire and move to France.

4.7   Shipman's Arrest and Conviction

It was Kathleen Grundy's daughter who brought Shipman to justice. She refused to believe that the new will was genuine and tracked down the two people who were supposed to have witnessed her mother's signature. After Shipman's conviction, detectives paid tribute to her determination and perseverance, without which there would almost certainly have been no trial. It was at her insistence that they decided to interview Shipman, who said that Kathleen Grundy had borrowed an old typewriter from him but he was not certain when that had been. He showed them the machine and they took it away. However, none of the fingerprints on it were those of Kathleen Grundy, and this was also true of the new will.

Exactly one month after she had been buried, Kathleen Grundy was disinterred. Samples of muscle and liver tissue were taken from the body and sent for analysis to the Forensic Science Unit at Chorley in central Lancashire. The toxicologist, Julie Evans, used mass spectrometry to identify and quantify the morphine present and confirmed that was enough there to have caused her death. The police now had sufficient evidence to arrest Shipman. He was indignant. He insisted that her medical records would show his innocence, because his suspicions of her drug misuse were recorded, and that two years previously he had begun to suspect that she was taking morphine, heroin or pethidine. In fact Shipman had forged these notes, but only after he heard that her body had been exhumed.

Shipman appeared before Tameside Magistrates Court on 8 September 1998 charged with the murder of Kathleen Grundy. At the same time Detective Superintendent Postles applied for three other exhumation warrants, and these were carried out later that month. The first body was that of Joan Melia, whose remains indicated the presence of diamorphine and showed that she had not died of pneumonia, as Shipman had claimed. Next was Winifred Mellor, who had been found dead at 6 pm on 11 May 1998 sitting in her favourite chair. The pathologist, Dr John Rutherford, noted a bruise on Winifred's arm which indicated that she had been given an injection directly into a vein. Then it was the turn of Bianka Pomfret's body to be exhumed. Bianka had died suddenly and was found by her son on 10 December 1997. When her body was disinterred nine months later there was no indication of death due to natural causes, and no abnormality of the heart or blood vessels. The toxicologist again found abnormally high levels of diamorphine in her body tissues.

By the time Shipman was interviewed by detectives on 7 October 1998 they knew that he had created false entries in his medical records, but they needed him to confirm that he was the only person who had entered patient's notes into his computer, which he did. Medical records are completely confidential and can be kept secret even from the police. Shipman knew this, and his lawyer successfully challenged the right of the police to seize them and as a result they were returned to his surgery—but it was too late to save him. He was unaware that his computer contained a record of the exact times when he had created back-dated entries for the previous two or three years showing that his patients had been suffering ill health compatible with their supposed cause of death. This was certainly true of Kathleen Grundy's records, making it appear that she had a serious heart condition and also that she was addicted to drugs. At 4 pm on the very day she died, Shipman had forged a false medical history for her. When told that his computer had logged the exact times he had done this Shipman broke down and wept in frustration. There and then he was charged with the murders of Winifred Mellor, Joan Melia and Bianka Pomfret.

The police had to establish how he had obtained the morphine and diamorphine. They consulted local pharmacists, who confirmed that Shipman had prescribed these drugs for a large number of his seriously ill patients and that he had been in the habit of collecting the completed prescriptions, saying he would deliver them himself. This he did, of course, although he invariably kept at least half for his own use. Shipman's car was examined some weeks after his arrest and was found to contain a large quantity of diamorphine.

Shipman claimed that the morphine and diamorphine found in the bodies of his patients arose from the fact that they were habitual users and had taken an overdose. This theory was disproved by a forensic hair analyst, Dr Hans Sachs, who examined strands of hair from the exhumed bodies. He was unable to find traces of these drugs in the hair, which clearly demonstrated that they were not drug addicts, as he claimed. In an addict the level of morphine would be around 2 nanograms per mg of hair, but in the hair of his victims there was none.

Meanwhile the police exhumed more bodies, and these had been buried for over a year. Marie Quinn had been 67 when she died the year previously, on 24 November 1997. The pathologist who examined her body found no indication of a stroke but again confirmed by analysis that she had died of a lethal dose of morphine. Ivy Lomas was 63 when she died unexpectedly in Shipman's surgery in May 1997. Two months later, on 11 July 1997, Irene Turner, aged 67, was found dead on her bed fully clothed. Muriel Grimshaw was 76 and died three days later, on 14 July 1997. The bodies of all three were found to have morphine levels indicating poisoning. The police worked backwards until they reached the point where the exhumed bodies were too decomposed to allow meaningful analysis. This was found to be the case with Alice Kitchen, who was murdered on 17 June 1994. Bodies buried prior to that date were left undisturbed.

The trial of Dr Harold Shipman began on 5 October 1999. He was charged with 15 murders and one of forgery. The murders were of nine women who had been buried (Kathleen Grundy, Winifred Mellow, Joan Melia, Bianka Pomfret, Marie Quinn, Ivy Lomas, Jean Lilley, Irene Turner and Muriel Grimshaw) and six who had been cremated (Norah Nuttall, Maureen Ward, Pamela Hillier, Maria West, Kathleen Wagstaff and Lizzie Adams). The forgery charge related to Kathleen Grundy's will.

The trial lasted six weeks, and 120 witnesses gave evidence. The judge was Mr Justice Forbes and Shipman's defence lawyer was Nicola Davies. On Monday 31 January 2000 the jury, having taken six days to reach their verdicts, announced that they found Shipman guilty of all the charges. The judge sentenced him to 15 consecutive life sentences.

The guilty verdict was not the end of the affair. The police suspected that there might have been at least 50 further murders committed by Shipman, some of which they had already investigated. Fresh inquests were held on some of these and the cause of death amended to unlawful killing. As more and more people realised that their own relatives might also have been murdered by Shipman, public pressure mounted for an inquiry, and on 21 September 2000 it was announced that this would be headed by the High Court Judge, Dame Janet Smith. She examined all the deaths which had occurred among Shipman's patients during his long career, 887 in all. Dame Janet spent the next three years working through each of them, taking 2300 witness statements and examining 360 sets of doctor's notes. Of the 887, there was compelling evidence that 394 of the deaths were natural and 493 were suspicious. Of the latter, the majority had occurred in the patient's own home, 121 had occurred in residential homes and 15 in Shipman's surgery. Applying stricter criteria, to decide whether any of these individuals had been murdered, she whittled the number of possible murders down to 298. She finally came to the conclusion that 215 of these had definitely been murdered, 45 were highly suspicious, and the remaining 38 were possibly murders but the evidence was incomplete.

Shipman also stole jewellery from some of his victims, and around 100 such items were found in a box in his garage. His wife, Primrose, was able to produce documentary evidence that some were hers, but she admitted that 33 items were not hers and these were handed over to the Asset Recovery Agency in May 2005. In the event there was just one item, a platinum diamond ring, for which the relatives of a woman Shipman had murdered were able to successfully provide photographic evidence in support of their claim of ownership. The remainder realised £1700 when sold at auction, and the proceeds were donated to the Tameside Victim Support Group. A Garden of Tranquility in Hyde Park was opened on 30 July 2005 as a memorial to Shipman's victims.

4.8   Postscript

On 13 January 2004 Shipman tore a bed sheet into strips, twisted these into a rope and hanged himself from the bars of his cell window. He was 58. Shipman's estate came to £24 000, which he left to Primrose in a will dating from 1979. He had ticked the 'cremation' box.

CHAPTER 5

Adrenaline and the Near-Perfect Murders of Kristen Gilbert

A word shown in **bold** *indicates that further information will be found in the Glossary*

5.1 One Way to Deal with Difficult Patients

Adrenaline is an activating life-saving chemical which the body produces in times of crisis. In addition doctors can inject it at times of emergency, for example to restart a heart which has stopped beating. Such is its power to influence vital bodily activities that adrenaline is effective at the microgram level. And yet the body can easily have too much of a good thing, in other words a large dose of adrenaline can so unbalance bodily functions, and especially the heartbeat, that it can cause death. That a natural chemical present in the body should also be able to act as a dangerous toxin makes it a perfect agent of attack.

In July 1995, Stanley Jagodowski, a grumpy, obese 66-year-old US army veteran, had a leg amputated due to gangrene. On the evening of 21 August, at the Veteran Affairs Medical Centre in Leeds, Massachusetts, he had been made comfortable for the night by two nurses, who then saw one of their colleagues, Kristen Gilbert, enter his room with a syringe in her hand. A few minutes later his alarm sounded when he went into sudden cardiac arrest. Despite the best efforts of the medical staff, three hours later he was dead.

Molecules of Murder: Criminal Molecules and Classic Cases
By John Emsley
© John Emsley, 2008

Kristen Gilbert thought she had found the perfect poison—ideal for disposing of yet another of those tiresome charges she was expected to deal with. But she was wrong. Three years later, on 13 July 1998, Jagodowski's remains were exhumed and tests showed the presence of an excessive level of adrenaline, far more than his body would have produced even in a life-threatening situation. In 2001 she was found guilty of murdering him and four other patients, and was suspected of killing a further 50 or more during the years she had been a nurse.

5.2   Adrenaline (Epinephrine)[i]

In 2003 I enjoyed an evening meal of baked salmon. The following morning I awoke with a curious numbness in the mouth, rather like that of a painkilling injection. As the hours passed, the numbness spread to my face and neck, and eventually I had to call a doctor. He diagnosed an allergic reaction to something I had eaten and gave me an injection of a powerful antihistamine, with an emergency number to phone if the numbness led to difficulty in breathing or swallowing. Thankfully this did not happen, but the next day I was provided with a spring-loaded adrenaline syringe and shown how to use it in the future if I needed to. In that case I would press the needle hard against my thigh for a large dose of adrenaline to be injected into my body, giving time to call medical help. A large dose is a relative term, and in this case it means 0.3 ml of a 0.1% solution, as little as 0.3 *micro*grams. Such is the powerful effect of this natural chemical.

Adrenaline was the first hormone to be identified as such. Its discovery began in 1856 with the work of a French physiologist, Charles-Edouard Brown-Séquard (1817–94), who showed that removing the tiny adrenal glands from the liver of an animal would quickly cause it to die. Clearly these glands were producing a substance essential to life—but what might it be? That same year his colleague, Alfred Vulpian (1826–87), demonstrated that whatever it was, it was carried from the adrenal glands by the blood. But he too was unable to say what it was. Extracts of the gland were known to be very potent at raising blood pressure, if only for a short time.

---

[i] In the USA the spelling is Adrenalin, but this is also a proprietary name owned by Parke–Davis & Company.

In the mid-1800s organic chemistry was still in its infancy, and the structure of a molecule such as adrenaline would not be fully understood for another 30 years. Two chemists who did the most to develop the idea of molecular structure were the 29-year-old August Kekulé (1829–96) in 1858, and 22-year-old Jacobus van't Hoff (1852–1911) in 1878. Kekulé developed the concept of chemical bonding, and van't Hoff proposed the three-dimensional view of molecular structure which led to the theory of **chirality**. The latter concept was ridiculed at the time—but van't Hoff had the last laugh when in 1901 he received the first ever Nobel Prize for Chemistry.

This background explains why it took almost 50 years for an organic molecule such as adrenaline, although not particularly complex by today's standards, to be understood and synthesised commercially. In May 1886 William Bates reported in the *New York Medical Journal* that it was a simple chemical which was being produced by the adrenal gland, and it was this substance which was having such a powerful effect.

Others were now spurred into identifying it and three chemists appear to have done so independently. In 1895 the Polish scientist Napoleon Cybulski (1854–1919) became the first to isolate it in a pure state. Two years later an American, John Jacob Abel (1857–1938), repeated this and in 1903 correctly reported its composition to be $C_9H_{13}NO_3$; he gave it the name **epinephrine**. Abel established the first department of pharmacology in the USA in 1893, at the newly founded Johns Hopkins University in Baltimore, Maryland, where he carried out his ground-breaking work on adrenaline. Meanwhile, Jokichi Takamine (1854–1922) was also working on the active agent from the adrenal gland, indeed there is some indication that he may have been the first to isolate it, but his paper describing this was not published until 1901. Takamine was born in Japan, graduated from the University of Tokyo, carried out postgraduate research at the University of Glasgow, and then returned to Japan in 1893. He married an American, Caroline Field Hitch, in 1884 and emigrated to the USA in 1894.

Parallel research on adrenaline was also conducted by pharmaceutical companies, with a view to its commercial sale. Initially they obtained it from the adrenal glands of animals, and it was sold in 1900 by Hoechst in the preparation Suprarenin. In 1901, Parke–Davis in the USA also began to sell their own version, named

Adrenalin, which is easy to confuse with the generic name, adrenaline. Clearly, if it could be produced synthetically it would become more consistently available and more reliable in quality. Friedrich Stolz at Hoechst eventually succeeded in synthesising adrenaline and patented his process in 1904, following which Hoechst began production in 1906. The adrenaline available nowadays is exclusively synthetic and is referred to under its chemical name, epinephrine. It is offered in a variety of commercial forms, including Ana–Guard, Anapen, EpiPen and Eppy.

Epinephrine may be a natural chemical but it is a toxic material, and has an **LD$_{50}$** of 4 mg per kg body weight. While it can be a life-saving drug, when it is injected into a person whose heart is frail it can kill, thanks to its ability to make the organ beat itself to destruction. This is exactly what nurse Kristen Gilbert knew would happen when she injected her victims. The antidote for epinephrine is an immediate intravenous injection of a beta-blocker such as propranolol, and there have been cases on record of people surviving massive overdoses of epinephrine, as happened to a 38-year-old man who was injected in error with 30 mg (30 000 micrograms).

Adrenaline is produced and stored in the adrenal glands, of which there is one at the upper end of each kidney, and its release is triggered by nerve impulses coming from the brain. The molecule is derived from either of two amino acids, phenylalanine or tyrosine. Fear causes the release of a large amount of adrenaline into the blood stream, where it quickly activates a variety of receptors.

Adrenaline acts as a neurotransmitter in the body, its action being to shift bodily processes up a gear. This effect is most noticeable in the heart, which immediately begins to pound, an indication both of increased heart rate and in the volume of blood pumped at each stroke. Adrenaline constricts the tiny blood vessels in the skin, while dilating those of the muscles and liver. In the liver it causes the release of glucose from its energy store so that this can quickly be transported to the muscles. It also dilates the pupils of the eyes and expands their field of vision.

Adrenaline in the blood stream dramatically increases during times of stress, such as when we are under threat, or when we have to speak in public, or if we unexpectedly receive bad news, or if we feel physically threatened. Adrenaline prepares us for whatever course of action may be required—whether to take flight or to stand and fight. Once the threat has passed the excess adrenaline quickly disappears and the heartbeat returns to normal. The body deactivates adrenaline by means of the action of *monoamine oxidase* enzymes, and discharges the degradation products *via* the urine.

Some people are known to look for situations which cause a rush of adrenaline; these are known as 'adrenaline junkies'. While there is no evidence that it is an addiction, these people experience a high from the pleasant result when it boosts certain brain chemicals, especially the endorphins, which also act as painkillers. This applies to activities which carry a high degree of risk and which will automatically increase adrenaline levels, for example skydiving, bungee jumping and hang-gliding.

Epinephrine has a number of medical applications, including the most life-threatening such as dealing with a heart attack. It is also effective in another emergency situation, anaphylactic shock, and it does this by suppressing the immune system which is in error overreacting to an assumed threat. Epinephrine can relieve an asthma attack by relaxing the airways in the lungs. It is also useful in other allergic responses, such as drug reactions, hives and hay fever. An over the counter epinephrine inhaler, Primatene, has been available in the USA for use by asthmatics since 1984. Epinephrine is also used as eye drops for reducing pressure within the eye prior to eye surgery. It can control bleeding, and prolongs the effectiveness of local anaesthetics.

Injected doses of epinephrine lie within the range 200–500 µg (0.2–0.5 mg). It is used as either a 0.1% solution, the strongest, or as a 0.01%, or even a 0.001% solution, the last of these being the level used in local anaesthetics such as in dentistry. A 0.01% solution is the concentration injected directly into the heart to restart it.

Epinephrine is used in conjunction with a local anaesthetic in tumescent liposuction, a cosmetic procedure to remove surplus body fat. A mixture of lidocaine and epinephrine is injected under the skin and this anaesthetises a large area, permitting the removal of a considerable volume of unwanted tissue. The procedure was developed in 1985 by the American dermatologist, Jeffrey Klein, who believed it would be possible to perform the operation under local anaesthetic. He experimented with a painkiller comprising of 500 mg of lidocaine and 1 mg of epinephrine in 50 ml water. The first patient on whom the operation was carried out reported a stinging pain and an increased heart rate. Klein tried a more dilute solution on the next patient and an even more dilute solution on subsequent patients, none of whom complained of pain during the operation. Eventually he showed that a concentration ten times weaker than the original was effective.

5.3 The Crimes of Kristen Gilbert[ii]

After the infamous witch trials at Salem in 1692, when 19 women were hanged, only male murderers have been executed in Massachusetts. However, in 2001 it appeared likely that a woman murderer would be condemned to death, not merely because Kristen Gilbert's crime was so horrendous, and even those opposed to capital punishment would feel it appropriate, but due to the fact that her crimes were carried out on state-owned (federal) territory. If guilty of first-degree murder, which is homicide with intent to kill, she could therefore be executed. While this adds an element of legal interest, it is the nature of her chosen poison which is our primary interest. She used adrenaline, knowing it to be a perfectly natural chemical produced by the human body—and which it quickly disposes of—and as such it would seem to be the perfect murder weapon. While adrenaline was her preferred poison, however, she sometimes resorted to potassium chloride or insulin, both of which are also naturally present in the body.

Kristen Gilbert's motives for murder were varied. Sometimes she simply disposed of a terminally ill patient who was a nuisance but more often she liked to create an emergency, after which she would receive praise for her undoubted nursing skills. But occasionally she killed merely as a means of spending more time with her lover.

Apart from her cavalier attitude to the lives of her patients, Kristen was a highly professional nurse who clearly enjoyed her job at the Veteran Affairs Medical Centre (VAMC). Within a year of arriving, her fellow workers noticed that there seemed to be a lot more patient crises when she was on duty, and that a number of these seemed to end in death. Initially this was just put down to Kristen's bad luck—it could happen to anybody—and the doctors and fellow nurses were grateful that she was there to help. But eventually 'Angel of Death' became her colleagues' nickname for her, because she always seemed to be involved in so many fatalities.

[ii] Although it is usual to refer to people by their surname, in this chapter I will use the murderer's first name, Kristen.

Why would anyone suspect this young, good-looking, hard-working wife, mother and nurse of being a serial killer? We will never know for certain how many people she murdered, but it was probably around 50. Her killings reached a crescendo in the months between October 1995 and February 1996, when she was dispatching an average of one patient every week. During this period it was no longer possible to turn a blind eye to the slaughter, at least not by the other nursing staff, even though the hospital administrators seemed blissfully unaware of what was going on.

Used as a murder weapon, a shot of epinephrine can kill a sick or elderly person, yet the presence of the drug in the body will probably go unnoticed at autopsy. Only specific testing for an unusually high level of epinephrine in the blood will indicate anything abnormal, but this of course might well have been injected by a doctor in an attempt to save the patient's life—which is why a supply of epinephrine is always on hand in case a doctor or paramedic might need it.

There seems to be little in the background of Kristen's life to explain why she became a serial poisoner. She was born on 13 November 1967, the first child of a teacher, Claudia Strickland, and her husband, Richard, a serviceman with the US Coast Guard, who was stationed off the coast of Hawaii when he received the news of her birth. Shortly after Kristen was born the family moved to Fall River, Massachusetts, infamous as the town where Sunday school teacher Lizzie Borden almost certainly murdered her parents with an axe in a most brutal manner—but that was back in 1892, and in any case she was found not guilty. Nonetheless it had made that obscure mill town world famous, and indeed after the Strickland family later moved to Groton, Connecticut, Kristen liked to shock her fellow pupils by claiming she was related to Lizzie Borden. Kristen was a bright child, and she learned to play the cornet at school and was a member of the marching band, the orchestra and the jazz group. She was also in the school's maths team. She graduated a year early, in 1984, with very high grades and at 16 enrolled at the Bridgewater State College on a premedical course.

In the summer of 1986 Kristen fell in love with Glenn Gilbert, and in order to be near him she transferred the following year to Wachusett Community College and then to Greenfield Community College. Kristen had a part-time job as a home health carer for the Visiting Nurses' Association of Franklin County. In January 1988 she and Glenn eloped and Kristen became Kristen Gilbert. Soon after the marriage Glenn uncovered a rather different side to his wife's nature. One day when they were having an argument in the kitchen she suddenly lunged at him with sharp knife. He fled into the bedroom and locked the door until she had calmed down. In most other respects she was a conventional wife. Their first child, Brian, was born on 15 November 1991, and their second son, Raymond, on Kristen's birthday in 1993.

Kristen graduated from Greenfield Community College with a nursing diploma. In March 1989 she obtained a position at the Veteran Affairs Medical Center (VAMC) in the Leeds area of Northampton, Massachusetts, a town with a population of 30 000, of mainly self-employed and professional people. Calvin Coolidge, President of the United States from 1924–28, had been mayor in the early 1900s, and the town is famous for Smith College, a prestigious liberal arts institution for women.

Kristen was an asset to the local community. Not only was she an efficient nurse and to the outside world a loving wife and mother, she also carried out voluntary work at a local shelter for the homeless and spent time sewing Christmas stockings and quilts for charity. She was also a fun person when it came to parties. Underneath this façade, however, there lurked a selfish, homicidal drama queen.

The abnormal incidence of sudden deaths which occurred when Kristen was on duty was first noticed in 1991. There were 31 such incidents during her first two years at the hospital, of which 22 were apparently discovered by Kristen herself, far more than any other nurse. This was put down to coincidence, yet these alarming incidents continued at the rate of around one a month.

During the summer of 1995 Kristen's marriage began to break down and she transferred her affections to James Perrault, a Gulf War veteran who had started work at the VAMC as a security guard. Perrault was tall, dark and handsome—and physically very fit. While his main role was to police the VAMC, he was nevertheless expected to be on hand if there was a medical emergency. These gave him the opportunity to meet and flirt with Kristen. By September they were regularly having sex, sometimes in the back of Kristen's car. In November 1995 Perrault rented a one-bedroomed apartment near to where the Gilberts lived, and in December Kristen left Glenn and her two young boys and moved into an apartment of her own in Northampton Street. (There was some suspicion that before she left her family she had attempted to poison Glenn with potassium chloride, but it merely made him very ill.)

The deaths on Ward C during 1994 and 1995 totalled more than 90, of which 23 took place on the night shift (midnight to 8:00 am), 22 during the day shift (8:00 am to 4:00 pm), and 50 on the evening shift (4:00 pm to midnight), when Kristen was on duty. During the years 1991–93 there had also been twice as many deaths during Kristen's day shift than during the night and the evening shifts. When Kristen was eventually brought to trial, the prosecuting counsel asked the Dean of the University of the Massachusetts School of Public Heath to calculate the probability that the extra deaths during Kristen's shifts were due to chance, and his estimate was one in a hundred million. (The judge at the trial did not allow this statistical evidence to be given, for reasons that are not entirely clear.) This disproportionate death rate had not been picked up by the hospital management, nor had anyone linked this high death rate to the delivery of 135 ampoules of epinephrine to the hospital, none of it ordered by the doctors.

From the time she started working at the VAMC in 1989 Kristen activated the emergency alarm 72 times, compared to a total of 70 for all the other nurses combined. In addition, Ward C suffered 11 fires during the years between 1990 and 1993 and Kristen had discovered eight of them, and had even received an award for putting one of them out herself. There was also a curious bomb scare in 1993. It appears that Kristen enjoyed the excitement of emergencies, and that she was in her element dealing with them.

When Kristen was finally brought to trial, it was the deaths immediately preceding her suspension from duty which constituted the case against her, and these are the ones for which there is clearest evidence of the method she used to attack her patients.

5.4 Murders and Attempted Murders

US Army veteran, Stanley Jagodowski, was 66 when he was admitted to the Leeds VAMC on 21 July 1995. He was suffering from diabetes and high blood pressure, and was recovering from a leg amputation after it had become gangrenous. Jagodowski was a smoker, heavy drinker and lover of fast food, and he was seriously overweight (228 pounds).[iii] Combined with his slight stature (5 ft 7 in; 1.7 m) this gave him a **body mass index** of 35, putting him well into the obese category. He and his wife Claire had been married for almost 40 years but she could no longer manage to look after him, and when he had recently fallen she called the police to lift him. In hospital he began to improve, although he continued to be uncooperative and argumentative. In August he was transferred to Ward C, one of the long-term nursing units.

On the evening of 21 August, Jagodowski had been settled down by two nurses who left him at 8:20 pm. A few minutes later they saw Kristen Gilbert enter Jagodowski's room with a syringe in her hand and heard him cry out, "Stop! Stop! You're killing me!" as she was giving him an injection, before immediately leaving the room. That he needed an injection puzzled the two nurses, because none had been indicated, and when they returned to see what was the matter they found Jagodowski holding his arm, clearly still in pain, but otherwise appeared to be fine. No sooner had they left him again, at 8:43 he went into sudden cardiac arrest and nurses and doctors rushed to his aid. Emergency resuscitation successfully restarted his heart and he was transferred into intensive care on a ventilator—but he died just before midnight.

Three years later, on 13 July 1998, his remains were exhumed and the presence of high levels of epinephrine was confirmed. Kristen had injected him with this drug, known to be extremely painful, which explains why Jagodowski cried out. It takes several minutes for the drug to reach the heart, where it causes a massive increase in activity. The delay allowed Kristen to be well away when the emergency alarm was sounded.

Henry Hudon was also murdered by Kristen, and his story is rather sad. His life was changed permanently when he became involved in a fight in a pizza restaurant in London, UK, in 1986 and he was struck in the face with a bottle. The blow was so severe that it knocked him out, smashed his front teeth and detached the retina of his right eye. Henry was 26 and a member of the US Air Force, and had been sightseeing in London with his wife and young children. He had merely intervened in a fight to try and stop it, but for the next three weeks he lay in a coma. When he finally recovered consciousness it was clear that he had suffered brain damage, which

---

[iii] This corresponds to $16\frac{1}{4}$ stones, or 102 kg.

eventually led to the diagnosis of paranoid schizophrenia. After returning home Hudon was put on permanent medication, but it did not always produce the desired effect and when this happened he would turn up at the Leeds VAMC for treatment. He sometimes remained there as long as a year before being discharged.

In 1995 Hudon had became unstable once again and he also developed a condition known as tardive dysdiadochokinesis, the most obvious symptom of which is an uncontrollable shaking of the body. His speech was slurred and he was sometimes verbally abusive without cause. He was admitted to the VAMC in November 1995, discharged himself on 7 December, but was re-admitted the following day following an overdose of medication which had made him very ill. He was put on intensive care in Ward C. That evening Kristen carried out her first attack on him with an injection of potassium chloride solution. We know that she chose this as her murder weapon rather than epinephrine, because instead of his heartbeat becoming increasingly rapid, it suddenly stopped beating altogether—which is precisely the effect of potassium chloride.

Kristen knew that epinephrine would not be likely to kill a 35-year-old, so she resorted to one of the chemicals used when criminals are executed in the USA: potassium chloride. She may also have read of its employment by a serial killer nurse in England, when trainee nurse Beverley Allitt killed the children in her care with it—see below.

Potassium: The Essential Element that can Kill

Over a period of ten weeks, from February to May 1991, at the Grantham and Kesteven General Hospital, Allitt injected 10 children with potassium chloride and four of them died. Her first victim was a seven-week old baby, Liam Taylor, and the last one a young girl, Claire Peck, who had been admitted to hospital suffering from a severe asthmatic attack. Allitt's colleagues had their suspicions that she was responsible for the sudden spate of emergencies and deaths, and these were confirmed when they asked for an autopsy on her latest victim. This showed an unnaturally high level of potassium. The hospital authorities called in the police and Allitt was arrested. She came to court in March 1993, and after a trial lasting nearly two months was found guilty and given multiple life sentences.

Why did Beverley Allitt behave as she did? The answer would appear to be that she was suffering from a rare mental state known as Munchausen syndrome by proxy. This condition was first recognised in 1977. Munchausen syndrome itself affects those who feign illness and seek medical help for symptoms they themselves have been responsible for; they can be quite convincing, and may well succeed in fooling doctors. On the other hand, Munchausen syndrome by proxy is the act of deliberately hurting someone else, often a person for whom one has a caring relationship, such as a parent hurting a child. Again the object is to seek medical help at the same time as appearing concerned about the victim's welfare.

Another nurse who murdered patients by injecting potassium chloride solution was the 36-year-old Orville Lynn Majors. He worked at the Vermillion County Hospital in Clinton, Indiana, and in November 1999 was found guilty of murdering six elderly people; he was sentenced to 360 years detention. He began work at the hospital in May 1993 and during the succeeding two years 130 patients died suddenly during his shift, twice as many as on other shifts. In the years before his arrival there had been around 25 deaths a year, but in 1994 the number increased to 101. There was one death for every 23 hours worked by Majors, but when he was not on duty the rate fell to one death every 551 hours.

On Sunday, 8 December 1995, Henry Hudon appeared to be recovering from his overdose and vomiting of the previous day, and his blood pressure and heart rate were back to normal. He was described as alert and comfortable at 5:30 pm, but 30 minutes later he was in serious trouble—his heart had suddenly stopped. The first doctor on the scene decided that there was no time to lose and ordered a massive dose of epinephrine, and although he appeared to have been dead for more than 20 minutes it eventually brought Henry back to life. Clearly Kristen had not used epinephrine to attack him, since the administration of a further quantity would hardly have saved his life. Not that it was saved for long, because his heart stopped a second time an hour later, and again was restarted, this time using an injection of bretylium, a drug used to control a highly irregular heartbeat. Then at 7:30 Hudon suffered a further heart attack and was again revived, but after a fourth, at 9:30, he was declared dead.

While attending to Hudon, Kristen had found time to e-mail her lover, Perrault, arranging to meet him after work around midnight.

Sometimes Kristen's attempts at murder failed, as in the case of Thomas Callahan. He was a 61-year-old who had served in the Korean War in the early 1950s. He was a heavy drinker (a bottle of whisky a day) and a heavy smoker (40 a day). He suffered from schizophrenia and was regularly admitted to Leeds VAMC to treat his chronic obstructive pulmonary disease (COPD), whose symptoms are breathlessness, wheezing and a persistent cough. He had in fact been admitted more than 20 times during the past 10 years. The latest was on 18 January 1996, and the nurses knew he was likely to be a difficult patient. However, there seemed to be nothing wrong with his heart, and his appetite was good. That evening it was Kristen's turn to look after him and she gave him an injection of epinephrine, which caused his heart rate to soar to 215 beats per minute and his blood pressure to rise to 191/116. However, he did not suffer the planned heart attack and within half an hour he was back to normal. Despite Kristen's care, Callahan had managed to survive.

Three of Kristen's colleagues, John Wall, Kathy Rix and Renee Walsh became convinced that she herself was responsible for the emergencies befalling Ward C, and after one of these they dis-covered three empty epinephrine vials in a disposal bucket. Not only that, but these had been the large dose variety used by doctors in cases of cardiac arrest, rather than the smaller vials used by nurses following a severe asthma attack or a case of food allergy. No doctor had authorised their use. Although they did not voice their suspicions at that time, they checked the unused epinephrine ampoules in the ward medicine cabinet—and there were 18 of them.

Kenny Cutting was another victim of Kristen's final frenzy of murder. He had married his high school sweetheart, Nancy, on 26 January 1976, and they lived in Lunenberg, Massachusetts. A year later he joined the US Army and he was still at the training camp at Fort Leonard Wood, Missouri, when their son Jeffrey was born on 11 April 1977. But Cutting's life and career were about to fall apart. After he had been in the army for a little over a year he was diagnosed as having multiple sclerosis. This incurable disease slowly eats away the nervous system, eventually reducing those it afflicts to a state of total dependency. Cutting had reached this state by the autumn of 1995, when he was transferred to the Leeds VAMC from the Jamaica Plains long-term care hospital to which he had been admitted in 1980.

By this time Cutting had already lost the use of his arms and legs. He went back to Jamaica Plains for an operation to correct a bowel obstruction, and on 1 November 1995 he returned to the VAMC to be treated with a course of antibiotics for an intestinal infection. On 26 January he was transferred to the intensive care unit in Ward C with a high fever. Within a few days this appeared to be subsiding, but there were still days when it was worse, and during these periods his heart rate increased.

On 2 February, before she left home to go to work, Kristen contacted her lover Perrault with plans for a night out. She said she would come round to his house at 10 pm, although that was two hours before her shift officially ended.

Throughout the day Cutting's condition improved—his blood pressure was a near-perfect 120/80, even though his fever remained high. Ward C had only 11 beds occupied, rather than its average occupancy rate of 25. When Kristen arrived for work she was assigned to the intensive care section, in which Cutting was the only occupant. During her first break she contacted supervisor John Wall and asked if she could leave early, adding somewhat surprisingly that Kenny Cutting was giving cause for concern and that he was likely to die. Soon after she returned to the intensive care unit, around 6:30 pm, Cutting took a turn for the worse and by 7:15 he was dead. With all the beds in the intensive care unit now empty, Kristen could leave work and was in her lover's arms soon after 10 pm.

Another survivor of one of Kristen's murderous attempts was Angelo Vella. He was a 68-year-old veteran of World War II. He too suffered from COPD and had spent many of his recent years at the Leeds VAMC; indeed he and his family were on friendly terms with Kristen. On 4 February he was back in Ward C complaining of shortness of breath. For some reason Kristen decided his days on earth were over, and soon after she arrived for work she was seen to inject him with something that made him cry out in pain so loudly that another nurse ran into the room to see what was amiss. She found Kristen holding a syringe and saw that Vella's heart rate was 300 beats a minute. A heart attack soon followed and the emergency alarm was raised. Such was the fear that Vella was about to die that a chaplain was called to administer the last rites, but his presence was not required since Vella recovered—and was actually able to give an account of what Kristen had done. The heart monitor had also recorded the events which had taken place, and the tapes were carefully collected and stored away.

The suspicions of Kristen's colleagues were now becoming acute, reinforcing what they had observed about the stock of epinephrine ampoules. On 6 December 1995 a box of 25 ampoules had been delivered to Ward C and these were clearly all used, since six weeks later another box of 25 ampoules was delivered—in fact, since August 1995 more than 100 ampoules of epinephrine had been used in Ward C. When Nurse Rix returned to work on 6 February she went to check the stock of epinephrine ampoules and found that only three remained.

Kristen's murderous career was finally drawing to a close, but not before she had dispatched Ed Skwira to the next world. He was born on 25 May 1927 and joined the US Army as a truck driver when he was 18. He left the army in 1947, married, and had three children. Ed was six feet tall and weighed 250 pounds.[iv] His BMI was 34, placing him in the severely obese category. He was a heavy drinker, so much so that when he was 68 years old he was being treated for alcohol abuse. In addition he suffered from diabetes and angina, which he controlled using nitroglycerine pills. He had also had abdominal surgery in the 1980s to repair his aortic artery.

When he arrived at the VAMC on 15 February 1996 Skwira was complaining of chest pains, and was sent to the intensive care unit in Ward C to be monitored. The doctor on duty, suspecting that his aorta might be leaking, ordered a chest X-ray and full-body CAT scan, which revealed a bulge in the artery which could be dangerous. He was also behaving strangely and appeared to be hallucinating. It was decided to transfer him in due course to the Baystate Medical Center for an operation, but before this could be arranged Kristen sounded the emergency alarm. Skwira was having some sort of attack, and a decision was immediately taken by the doctor in charge to send him to Baystate Hospital. Kristen was to accompany him.

With Kristen gone, Nurse Rix now went into Skwira's room and purposefully looked in the medicine cabinet to see whether the

---

[iv] Equivalent to $17\frac{1}{2}$ stones, or 112 kg.

three ampoules of epinephrine were still there—but they had gone. In the needle disposal bucket by Skwira's bed were the three empty ampoules. Rix collected the print-outs from the heart monitoring equipment, but when she looked for these the following morning they had disappeared.

Meanwhile, at the Baystate Medical Center the doctors quickly ruled out a heart attack, as Skwira's heart was functioning normally. They deduced that he had developed tears in his aorta and stomach wall and that there was little they could do to save his life. He died on 18 February 1996. When his body was exhumed nine months later the toxicologist discovered traces of the drug ketamine, which would have explained his odd behaviour at the VAMC. But this drug was never stocked at the hospital. (Ketamine can be used as an anaesthetic, but is better known as a street drug and goes under a variety of names, including kit-kat and special K.)

By now Kristen sensed that she was under suspicion and on 17 February 1996 she became hysterical, claiming that a patient had attacked her. She was allowed to go home, although at this stage no one was prepared to believe her story. She was told not to return and that investigators were being brought in to examine the hospital records. In the four months following Kristen's indefinite medical leave there were only four deaths on Ward C and two medical emergencies. In the four months before she left there had been 23 deaths and 30 emergencies.

5.5   Kristen Discovers Adrenaline is not the Perfect Poison

The Chief of Security at Leeds VAMC called the US Veterans' Administration Inspector General's Office to inform them that they thought a nurse might have been murdering patients, and Special Agent Steve Plante was sent to investigate. Rix and Wall explained their suspicions to Plante and suggested that he examine the patients' ECG charts[v] to confirm that they had been injected with epinephrine. When Plante was given the medical files of those who had died the previous year he found that this vital piece of evidence was almost always missing.

---

[v] ECG is short for electrocardiogram. In the USA it is known as EKG, short for electrokardiogramm, the German terminology and spelling.

Kristen probably learned from Perrault that she was being actively investigated, and she tried to convince him, and others that she knew, that she was being framed by jealous colleagues. Maybe Perrault initially believed her. Meanwhile she disrupted the working of the hospital by causing a bomb scare, leading to evacuation of the patients while the buildings were searched. On 9 July, Kristen went to Perrault's apartment, but behaved so irrationally that he called the police. They said she could either go to jail or get medical help. She elected to go for psychiatric treatment at Holyoake Hospital. Kristen continued to pester Perrault with phone calls, during one of which she admitted killing patients. Eventually he obtained a restraining order against her.

A Grand Jury[vi] met on Tuesday, 16 July 1996, and heard how Kristen had confessed to Perrault that she had killed patients with injections of epinephrine, and that she had told him that the drug would be impossible to trace after death due to its rapid breakdown in the body. They also heard from Special Agent Plante how 75 ampoules of epinephrine had been missing from the VAMC between 30 August 1995 and 21 February 1996. While the Grand Jury deliberated, Kristen remained at home, apparently attempting suicide with what she said was an overdose of aspirin. She was rushed to hospital, but was of course lying.

Kristen supposedly made a further attempt on her life in August 1996 when she took an overdose of a different painkiller and was readmitted to a local hospital for psychiatric care. The following month, and just after her discharge from hospital, she purchased a child's toy which could record and distort the human voice and used this to make threatening phone calls to Perrault and the VAMC in an attempt to divert attention from herself. She was back in the psychiatric ward on 2 October. A week later she was discharged but immediately rearrested, and on 15 October was tagged and released into the care of her parents, who lived on Long Island, the bail condition being that she did not return to Massachusetts or phone any potential witnesses. On 2 November she was charged with making false bomb threats and phone calls and placed under house arrest at her parents' home.

[vi] A Grand Jury decides whether the case against an accused person is sufficient for them to stand trial.

Kristen was initially brought to court and charged with making a malicious bomb threat at the VAMC. Her trial began on 7 January 1998, and she was found guilty and sentenced to 15 months' imprisonment in Danbury Federal Prison. Meanwhile a murder charge was being prepared. After serving 10 months of her sentence, Kristen was removed to the Hamden County House of Correction for Women to await her next trial.

The trial of Kristen Gilbert for murder began in November 2000 before a jury of nine women and three men. The presiding judge was Michael Ponsor. More than 70 witnesses were called and 200 pieces of evidence produced. Kristen was charged with the murder of Henry Hudon, Ed Skwira, and Kenny Cutting, and the attempted murder of Angelo Vella and Thomas Callahan. Six months later an additional charge of murder, that of Stanley Jagodowski, was added to the indictment. While in jail awaiting trial Kristen had secretly begun to write a novel, clearly based on her life. This was discovered and confiscated, but it could not be used in evidence against her. Nor did Judge Ponsor allow the prosecution to call evidence that Kristen had tried to murder her husband, Glenn, with an injection of potassium chloride. But he did allow Glenn to tell the court of her confession to murder during her phone call. Glenn also provided the investigators with one of Kristen's books which she had tried to reclaim from her former home. The book was the *Handbook of Poisoning*, and the pages dealing with epinephrine were folded down for easy reference. Dr Fredric Reiders, laboratory director of the National Medical Services in Pennsylvania, gave evidence that the bodies of Skwira, Jagodowski, Cutting and Hudon had significantly higher than normal levels of epinephrine in their tissues.

The jury began their deliberations on the 23 February 2001. After debating for almost three weeks, they delivered their verdict on 14 March: Kristen was guilty of the first-degree murder of Skwira, Cutting, and Hudon, and the second-degree murder of Jagodowski. As regards Vella and Callahan, she was found guilty of attempted murder. The question now was whether she would be executed. The jury were unable to reach a unanimous decision about this and she was therefore spared the death penalty. Judge Ponsor sentenced her to four consecutive life terms, plus 20 years, and without the possibility of parole. She is at present serving this sentence in the Federal Medical Center prison in Carswell, Texas.

Could she have been executed? The answer is yes, although the State of Massachusetts is not in favour of the death penalty. The last time a criminal was executed in Massachusetts was on 9 May 1947, when two men, 34-year-old Edward Gertson and 32-year-old-old Phillip Billeno, were electrocuted for the murder of Robert Williams. But Kristen's murders had been carried out on state property, so she was in effect subject to federal law, which included execution as the punishment for first-degree murder. Kristen sensibly dropped her appeal for a new trial when she discovered that a recent Supreme Court ruling would have allowed the prosecution to call for the death penalty.

Part II:
Dangerous Man-made Chemicals:
So Simple, So Useful, So Deadly

CHAPTER 6

Chloroform and Mrs Bartlett

A word shown in **bold** *indicates that further information will be found in the Glossary*

6.1 How did She Do It?

After Adelaide Bartlett had been found not guilty of poisoning her husband Edwin with **chloroform**, the eminent surgeon Sir James Paget said that in the interests of science she should reveal the method she had used. Of course she never did, but it is possible to work it out, as we shall see. Sir James had accepted the theory advanced by Adelaide's defence counsel that no one could drink chloroform and not realise how dangerous it was, due to the searing pain in the mouth and throat. Yet the autopsy and forensic examination of Edwin Bartlett's body showed the presence of liquid chloroform in his stomach, despite the fact that he appeared to have died peacefully on New Year's Eve, 1885. The circumstantial evidence against Adelaide was compelling: she had the poison, a strong motive for wanting him dead, and the perfect opportunity to murder him—yet she walked free from court.

Although one can no longer buy neat chloroform in a pharmacy it is still available from chemical suppliers, indeed it was used in the murder of a young girl as recently as 2005. So what is this curious chemical which Queen Victoria regarded as heaven-sent, but which doctors realised could cause certain death? And why was it Adelaide Bartlett's weapon of choice?

Molecules of Murder: Criminal Molecules and Classic Cases
By John Emsley
© John Emsley, 2008

6.2 Chloroform

Extreme pain is hard to bear, and when we find ourselves suffering it we expect a doctor to provide rapid relief. Yet for earlier generations it was often the physician himself who inflicted the greatest pain, at the time he was performing an operation. Before the days of anaesthetics surgeons had to work extremely quickly. Some were so skilled they could amputate a leg in 30 seconds, and even remove a bladder stone in under a minute. Meanwhile, all their patients could do was grin and bear it, and try not to scream in agony. Today we expect even major operations to be pain-free, and we may lie unconscious and relaxed for several hours while deep surgery is performed within our body. This state of blissful oblivion has been made possible thanks to safe and effective anaesthetics developed during the second half of the twentieth century.

Throughout recorded history doctors have been concerned about reducing the pain of surgery. Almost 2000 years ago the Greek physician Galen (129–216 AD) gave those on whom he operated a concoction of opium and mandragora (mandrake), but in this he seems to have been an exception. Around the same time but on the other side of the world, the Chinese physician Hua T'o performed operations on patients rendered drowsy by a tincture of cannabis resin. In his book, *De Materia Medica*, the Greek pharmacologist Dioscorides (40–90 AD) recommended that mandragora should be taken before surgery as this too can induce sleep, although not without some worrying side effects, as we saw in Chapter 2.

The first man-made anaesthetic was ether, chemically **diethyl ether**, made by reacting ethanol and sulfuric acid. Who discovered it we do not know, but it was experimented with in 1540 by the great physician and pharmacist, Paracelsus (1493–1541), who mixed it with bird feed and then observed that his hens quickly fell asleep, remained so for a long time, but then woke up apparently none the worse for the experience. This observation was not acted upon for another 300 years until on 30 March 1842 an American surgeon, Crawford Williamson Long, used it to put to sleep James Venable, from whose neck he then removed two tumours. His patient said afterwards that he had felt no pain, and gladly paid Long his two-dollar fee. Another American, William Clarke, of Rochester, New York, had also experimented with ether a few weeks earlier and had anaesthetised a Miss Hobbie before extracting a tooth.

Ether soon established itself as a way of making operations painless, although some surgeons resisted the use of anaesthetics as a matter of principle. Pain, they said, was natural and an essential part of the healing process. Chloroform, and Queen Victoria, were to change their minds.

Success has many parents, failure has none. Once chloroform became widely used in surgery, three people later independently claimed to have been the first to discover it in 1831. However, it was to be a further 16 years before it was used as an anaesthetic, and only many years later that people became interested to know who might have been the first to make it. The great German chemist Justus von Liebig claimed in a paper published in 1872 to have been its discoverer. The French, however, reserved this honour for the pharmacist Eugene Soubeiran. The truth was that neither man was the first—its true parent was an American, Dr Samuel Guthrie (1782–1848), living on the shores of Lake Ontario. He ran a variety of commercial enterprises, including a distillery and a gunpowder mill, but he also had a private chemical laboratory. It was there that he first produced pure chloroform. In fact he was trying to make 'chloric ether', which we now know as **1,2-dichloroethane**. This was a product of the reaction of **chlorine** and **ethene**, and appeared to have potential as a solvent. Guthrie thought it might be possible to produce it by heating whisky with bleaching powder, and he thought he had been successful since he obtained a liquid with a similar sweetish odour—but he had in fact made chloroform (trichoromethane). The date of his discovery was 1 July 1831.

None of these men realised that chloroform could be a powerful anaesthetic, and when a few years later it was included in the *United States Dispensatory* it was as a treatment for asthma, persistent coughs and other conditions in which breathing was difficult. The medicine was a 12% solution of chloroform in alcohol. Drinking this was effective, and it presented no obvious danger. It was later noticed that inhaling the vapour of chloroform produced a very different effect in the body: at first drowsiness, and then a deep sleep.

Chloroform was first tested as an anaesthetic in 1847 by Pierre Flourens using animals (1794–1867). Its use on humans began in the UK, where dentists in London were looking for an alternative to ether. In May of that year Michael Cudmore Furnell, a 17-year-old medical student at the Middlesex Hospital in London, had been forbidden access to ether because he kept getting high by sniffing it, so he carried out some experiments on himself using chloroform instead. The results were more spectacular, and they didn't leave him with the irritating cough which ether usually produced. He mentioned this to Jacob Bell, the pharmacist at the hospital, and Bell passed on the information to his fellow pharmacists.

The man whose name is most associated with popularising chloroform as a general anaesthetic was James Simpson (1811–1870), Professor of Obstetrics in the University of Edinburgh. He was acutely aware of the pain suffered by women in labour, and had used ether on 19 January 1847 to relieve the agony of a woman with a deformed pelvis. This was successful, and against all expectations she survived, although the baby unfortunately died. Later that year Simpson investigated other volatile chemicals as anaesthetics and concluded that there were three which had the desired effect: ethyl nitrate, benzene and carbon disulfide. None of these was suitable for use in an operating theatre, however, least of all carbon disulfide, which smells of rotting cabbage and is a neurotoxin. Later that year he was advised to try chloroform by a friend, David Waldie, a pharmacist at the Apothecaries' Hall in Liverpool, and on 4 November he held a dinner party for family and friends during which he invited those present to sniff chloroform. Most of them, including Simpson himself, ended up under the table. He had found what he was looking for.

So it was on 8 November 1847 that Jane Carstairs, whose first pregnancy had ended with a stillborn baby after a three-day ordeal of intense pain, was safely delivered of her second baby while under the influence of chloroform. Simpson immediately dashed off an article to the *Lancet*, which appeared in the 21 November edition, and his pamphlet on the subject was advertised in the local newspaper, *The Scotsman*. Within a week 1500 copies of the latter had been sold at 6 pence a copy. In it Simpson recommended that 1 to 3 drams[i] of chloroform, or up to 100 drops, should be allowed to drip on to a silk handkerchief over the patient's face. The extent of the anaesthesia could be judged from the sound of the patient's breathing.

Within a year almost all operations, at least those in Edinburgh, were performed on patients anaesthetised with chloroform, and by

---

[i] In the metric system this corresponds to 3–12 g, equivalent to 2–8 ml.

the end of 1848 it was being tried all over the world. Some surgeons still preferred to operate with the patient conscious, however, and there were also doctors who opposed its use in midwifery on the grounds that absence of labour pains put the mother's life at risk, since these were a useful guide to the progress of the delivery. Some doctors justified their opposition to chloroform by claiming that it led to pneumonia, insanity and general paralysis. One London doctor, George Thompson Gream, was even convinced that it gave rise to something far worse: erotic fantasies! Women under its influence were indeed sometimes believed to behave in an amorous, even a wanton, fashion.

6.3 But is Chloroform Safe?

Far more worrying to doctors as a whole were reports that some patients had died suddenly under the influence of chloroform. An early example of an inexplicable death occurred on Friday, 28 January 1848, when Hannah Greener of Newcastle-upon-Tyne was being operated on for an ingrowing nail on the big toe of her right foot. The corresponding nail on her left foot had been removed a few months earlier, with ether as anaesthetic. The family doctor now decided to use chloroform, but within a few minutes her breathing became rapid, then suddenly stopped. She was dead—the chloroform had killed her. Or had it? So began a controversy about the safety of this anaesthetic that was to rage for more than 50 years.

As we shall see, such sudden deaths occasionally occurred in England, but were rare in Scotland, India or France. And before it could be proved that chloroform was indeed a risk to life the head of the world's greatest empire, Queen Victoria, decided to try it—under the care of the world's most experienced anaesthetist, Dr John Snow (1813–1858). When she consulted him he was 40 years old and an advocate of healthy living—he was a vegetarian and a total abstainer from alcohol.[ii] Snow is best remembered for removing the handle of the Broad Street pump in 1853, thereby bringing to an end the cholera epidemic which had been the scourge of the Soho area of central London. His epidemiological observations linked the incidence of the disease to water from the

---

[ii] His 'healthy' lifestyle brought him no lasting benefit. He died of a stroke aged 45, soon after completing his book *On Chloroform and Other Anaesthetics*.

pump, which was contaminated by faeces. This was before doctors fully realised the nature of disease pathogens, the importance of which only became clear following the work of Louis Pasteur (1822–1895) some 25 years later.

Snow had previously used ether when called upon to anaesthetise patients for operations, tooth extractions or childbirth, but he was converted to the use of chloroform. He advised only light anaesthesia for women in labour, because he observed that babies born under its influence were less vigorous during the crucial early minutes of life. Snow recognised five states of chloroform anaesthesia: in the first and lightest, the patient remained conscious but elated as if drunk; during the second state the patient was still conscious but in a dream-like trance, which was the one he preferred in the case of a woman in labour; in the third state the patient was unconscious, relaxed, and no longer responsive to pain, the preferred state for surgery; in the fourth stage the patient was unconscious and breathing deeply, which he advocated only for the most difficult operations. In the fifth state, which he studied on animals, breathing became strained as the chest muscles became partially paralysed. Snow was aware that there was a fine line between states three, four and five. The third state was produced with about 1 ml of chloroform (25 drops), whereas double this amount could result in the fourth or even the fifth state. How much chloroform a patient actually inhaled was uncertain, given that most doctors used the silk handkerchief technique and removed it only when the patient appeared to be unconscious to the right degree. Snow was aware that patients sometimes died without explanation, but he attributed this to the careless use of chloroform.

Queen Victoria was a mother many times over. Her first child, also named Victoria, was born in 1840 when she was only 21. Her second, Edward, was born in 1841, becoming the Prince of Wales and later King Edward VII. Then followed Alice (1843), Alfred (1844), Helena (1846), Louise (1848), Arthur (1850), Leopold (1853), and finally Beatrice (1857). Perhaps it was hardly surprising that the Queen regarded childbearing as a burden, and in 1858 she wrote to her eldest daughter: "What you say of the pride of giving life to an immortal soul is very fine, dear. I think much more of our being like a cow or a dog at such moments; when our poor nature becomes so very animal."

In 1853 Queen Victoria was pregnant yet again, but relief was at hand. One of her ladies-in-waiting had had a tooth extracted under chloroform and reported back to the Queen how painless the experience had been. The Queen was determined to try chloroform and Snow duly obliged on Tuesday, 7 April. At no time did the Queen lose consciousness, and Prince Leopold was born at 1:13 pm that afternoon. The Queen was delighted. "Dr Snow gave that blessed chloroform and the effect was soothing, quieting, and delightful beyond measure", she wrote in her diary on 22 April. The Queen's decision overcame all opposition to the use of chloroform, and childbirth *à la reine* became the norm. When the Queen's final baby, Princess Beatrice, was born on 14 April 1857, the chloroform was administered by her husband, Prince Albert, himself.

While chloroform continued to be widely used, there was still a small dark cloud on the horizon concerning the unexplained deaths, and the anxiety continued to grow. Unexpected and inexplicable deaths continued to be reported in the media, especially when the person concerned happened to be young and strong. In 1864 a 26-year-old carpenter was to have a damaged bone removed from his foot but died even while he was requesting the anaesthetist to use less chloroform. The following year a young woman in Edinburgh died in the dentist's chair, yet he said she had assured him that she had had chloroform during a previous tooth extraction. In 1869 a 19-year-old student at Oxford died on the operating table, and the same year a 25-year-old pharmacist died within three minutes of being given chloroform. Doctors knew these to be rare occurrences, but they were nevertheless worrying. Was too much chloroform the problem? Did this have the effect of excluding air from the lungs? This possibility was reduced by dripping the chloroform on to a dome-shaped muslin mask placed over the patient's nose and mouth, ensuring that plenty of air was inhaled. An extensive survey of doctors by the Royal Medical Society was unable to reach any conclusion as to how these deaths were caused.

Accidents continued, in some cases even killing the doctor who had administered the chloroform. A Dr Adams at Clyde Street Hospital in Glasgow inhaled it in order to test its strength, found it somewhat weak and inhaled a second lungful, following which he collapsed and died. That was in 1849. A Dr Renwick of Alloa suffered a similar fate in 1860. At a hospital in the Westphalia region of Germany, an operation to remove a bullet lodged deep inside a patient took rather longer than planned, so more chloroform had to be administered, and yet more again. The air in the theatre became so heavy with the vapour that two surgeons and two nurses passed out. Unhappily, one of the nurses failed to recover and died two days later.

In 1870, Benjamin Richardson analysed 80 000 operations in which chloroform had been administered and assessed the risk of a patient dying as a result of the anaesthetic as 1 in 2500. A separate analysis of operations in the UK and the USA indicated that the death rate from chloroform was 1 in 2900, whereas for ether it was as low as 1 in 23 000. No matter how carefully it was administered the risk of death from chloroform remained unchanged. It was not the result of too much chloroform or too little air, but more probably due to the genetic make-up of the patient.

Thomas Brunton, following research into the effect of drugs on the heart, concluded that chloroform blocked the vagus nerve, which runs from the brain to the major organs, including the heart. Some surgeons blamed the deaths on impurities present in the chloroform, or to its decomposition products, and all of them insisted on absolute purity. From the earliest days it had been observed that chloroform slowly became acidic, and in 1882 it was shown that chloroform and air could react under the influence of light to form phosgene. This chemical is so toxic that in World War I it was used as a chemical warfare agent. Chloroform was thereafter always stored in dark tinted bottles to prevent access of light, and a special test was devised to check for the presence of phosgene. But these precautions still failed to prevent the occasional inexplicable death.

When chloroform was administered it posed two risks. Too much given too quickly could stop the heart immediately. Less chloroform over an extended period could paralyse the lungs, even though the heart continued to beat. Victorian doctors took a long time to recognise this double threat. Chloroform is able to depress the region of the brain which controls breathing and heartbeat, but it can actually paralyse the heart muscle itself.

As the number of cases of accidental death due to chloroform steadily mounted there was a move to return to ether as the anaesthetic of choice, and the campaign for this was led by Dr Benjamin Jeffries of Boston, Massachusetts. He came to London in August

1872, ostensibly to attend an international conference of ophthal-mologists, but he took the opportunity to promote the use of ether as the preferred anaesthetic. Jeffries read a paper to the conference emphasising the safety of ether and assuring his listeners that this anaesthetic had never caused a fatality. After the conference Jeffries visited 17 London hospitals to demonstrate its use, and by the time he returned to the USA there was a definite trend among younger London surgeons to regard ether as the better choice.

The British Medical Journal soon noted that more operations were being carried out under ether than under chloroform. By 1875, in London only Guy's Hospital was still using chloroform, as did all hospitals in Scotland, where it was widely believed that Scottish-produced chloroform was in some way superior and therefore safer. The great surgeon, Joseph Lister, founder of anti-septic surgery, continued to insist that chloroform was intrinsically safe, but even he witnessed its dangers at first-hand when a 27-year-old man from Manchester, who was having a deep abscess lanced, died on the operating table even before Lister had made the first incision.

In 1857, after the Crimean War, French army surgeons reported that in more than 30 000 instances in which the anaesthetic had been administered not a single accidental death had occurred due to its use. Chloroform retained its popularity in France, although at the Sorbonne, in Paris, the Professor of Physiology, Paul Bert (1833–1886), noted that the margin of safety for chloroform was probably lower than many doctors realised. He estimated the fatal dose to be only twice the anaesthetic dose, whereas for ether the fatal dose was proportionally much greater. Bert carried out ex-periments on dogs and confirmed how dangerous chloroform could be. One dog placed in an atmosphere containing 4 g of chloroform in 100 litres of air never became entirely unconscious or insensitive to pain, but died 10 hours later. When he increased the quantity to 6 g in the same volume of air a similar effect was observed, but death occurred after 7 hours. At 10 g, the dog became unconscious within a few minutes and died in 3 hours, and with 15 g it passed out immediately and died within 40 minutes. At 30 g it took the dog only a few breaths to become unconscious and the animal was dead in 3 minutes.

Chloroform gradually became less popular, and some medical bodies recommended its use should be discontinued completely; the

American Medical Association took this course of action in 1912. It continued to be used in the United Kingdom, indeed the highest number of chloroform fatalities occurred in 1933 when 278 patients died. An analysis of chloroform deaths before this showed that there had been 24 278 in total. On the global scale it was judged that in excess of 100 000 people had lost their lives to chloroform. (It could on the other hand be argued on a similar basis that that if an average of one person in 3000 had been killed by it, then there were 300 million operations in which the patient had safely bene-fited from its use.) What chloroform offered was a cheap, easily stored, readily portable anaesthetic, capable of relieving a great deal of suffering.

Doctors tried to address the problem of sudden death in two ways: some tried to find devices able to deliver chloroform at a more controlled rate, and others accepted the need for a safer anaesthetic. A surgeon in Leeds, UK, Thomas Nunneley, explored the use of combinations of volatile liquids, and developed ACE, a 1 : 2 : 3 mixture of alcohol, chloroform and ether, which became widely adopted. Chloroform continued to be used as an anaesthetic in some countries for more than 100 years, and it was still in use in parts of Africa in the 1980s.

6.4   Non-Anaesthetic Medical Applications

The use of chloroform in general medicine was much safer, and as late as the 1970s it was prescribed in doses up to 0.3 ml to calm the gut and reduce intestinal gases. In Victorian times a variety of chloroform-containing medicines became available, including mixtures of chloroform and morphine, either as a linctus (on a syrup base) or as a tincture (a solution in alcohol). At the present time the UK cough medicine, pholcodine, still contains chloroform and is highly effective for suppressing a 'tickly' cough.[iii]

Chloroform is metabolised in the liver to secondary products which may damage that organ. In 1976 it was banned in the USA by the FDA from use in cosmetics, drugs and food packaging materials. This was the result of it being a suspected human car-cinogen. Its carcinogeneity had been demonstrated in laboratory

---

[iii] These are unproductive coughs which may suddenly be triggered without any obvious cause.

mice, although this was never proved in humans. It was the toxicity of chloroform towards the liver which explained why some patients developed jaundice due to liver failure and died a few days later, despite recovering from the operation itself. A post-mortem tended to reveal damage to the liver in which cells had died and been replaced by fatty tissue, a symptom known as 'acute yellow atrophy of the liver'. Chloroform and its metabolites also caused damage to the kidneys and heart muscle, but it continued to be used as an anaesthetic since, unlike ether, it did not irritate the lungs—in fact in some patients the use of ether led to irritant pneumonia, although this was much rarer as a cause of post-operative death.

Chloroform pollutes the planet, or so it is claimed by those who oppose the chlorination of water supplies. There is no doubt that when water is chlorinated in order to kill off disease pathogens such as cholera, typhoid, meningitis and dysentery, then a trace of chloroform is produced by the reaction between the hypochlorite ion (ClO^-) and organic matter. However, the amounts are hardly detectable, and even the highly cautious US Environmental Protection Agency estimates that levels up to 0.3 ppm offer no risk—not that chloroform in drinking water ever reaches anywhere near this level.

There was a time when any new chemical was hailed as a potential cure for something, if only that something could be found, and chloroform was no exception. For example, it was discovered that toothache could be cured by a plug of cotton wool soaked in chloroform. In the USA chloroform became an in-gredient of cough syrups, ointments and even toothpaste. In the UK the biggest use of chloroform was in Dr Collis Browne's Chlorodyne, which first appeared in 1855 and whose composition was a closely guarded secret. This product was spectacularly suc-cessful and a supposed cure for many common ailments. The name was a contraction of 'chloroform anodyne', and in addition to chloroform it was believed to contain opium, capsicum, hydro-cyanic acid, ether and peppermint essence—or at least this was the recipe published in the British Medical Journal in 1862. In fact a few per cent of chloroform was added to many medicines as a sweetening agent and also for conditions where the patient was suffering from involuntary reflex reactions such as in tetanus, strychnine poisoning, convulsive coughs, and even rabies. Violent patients in lunatic asylums were forcibly given large doses to calm them down.

Cholera, the scourge of many British cities in the 1800s, caused painful stomach cramps, violent vomiting and excessive diarrhoea. These effects could to some extent be relieved using chloroform, allowing the body to retain liquid and nourishment—or to die in relative peace if severe dehydration could not be reversed.

6.5 The Abuse of Chloroform

Some individuals used chloroform so regularly that they were thought to became addicted. An example was James Bowskill, of Halifax, Yorkshire, who in 1892 was found unconscious and charged with attempted suicide, at that time a criminal offence. When he appeared before magistrates he pleaded addiction to chloroform, claiming that he habitually inhaled more than 50 ml daily; this was accepted by the court in mitigation. Clearly it was possible to remain on a high from chloroform if one sniffed it frequently enough.

The myth that a whiff of chloroform could instantly render someone unconsciousness was at one time widely believed. Many novels of the 1800s had villains rendering their victims helpless before robbing, kidnapping or 'ravishing' them. Victims of the last of these (rape) often claimed as much in court, despite medical evidence that chloroform was hardly ever used in this crime. The most likely cause of rape was that the victim was too drunk to resist, a factor with which the courts at that time had little sympathy. Even as late as 1975 nonsense stories about chloroform were still making news, and an item in *The Sunday Times* reported that passengers on the Orient Express had been robbed by opportunists who had sprayed chloroform into sleeping compartments, rendering the occupants unconscious.

Nevertheless, doctors and dentists were sometimes known to commit acts of indecency on women they had chloroformed. When the victim later complained she would often be told that women under the influence of chloroform tended to experience erotic dreams, and further action was not usually taken. Sometimes, however, the perpetrator was brought to justice, as in the case of the 33-year-old surgeon, Richard Freeman, in June 1870. He was asked to examine 24-year-old Lucy Ashby, whose father had taken her to the surgeon after he suspected his daughter of having sex with a young man who clearly had no intention of marrying her. Lucy maintained she was a virgin and the surgeon was asked to confirm this. The examination took a surprisingly long time, but Freeman finally confirmed that she was a virgin. She complained, however, that he had anaesthetised her with chloroform and when she came round she discovered that her clothes were in disarray and the surgeon had clearly been embarrassed. Her father immediately took her to a second doctor, who confirmed that she had been sexually assaulted. Freeman was arrested and sent for trial but he was acquitted, although the judge said that he must never again examine a female patient without another person present. Freeman was lucky, because he had been tried and acquitted of a similar offence three years earlier.

Suicide by chloroform was not unknown, but since it had to be breathed for some time to be effective, death was not easy to achieve. One doctor did succeed, but only by taping a bladder containing the liquid over his nose and mouth. Drinking it was even less successful, and would-be suicides usually did little more than damage their stomach lining and die a painful death some time later. As a method of murdering someone chloroform was far from ideal, but both as a vapour and in liquid form it has been used to this end.

6.6 Chloroform as a Murder Weapon

Life is full of strange coincidences. While writing this book I was surprised to read of a murder in a local Bedford newspaper in which the agent of death was chloroform.

Irchester is a small town in Northamptonshire where lived a married couple, 31-year-old Joanne Hall, a nurse at Bedford Hospital, and 33-year-old Gavin Hall, who worked as a radiographer at Northampton Hospital. They had two little girls, Milly (3) and baby Lucy (1). In 2005 the Halls had been married for six years but the relationship was in difficulty. Joanne was having a secret affair with a lawyer she met through a lonely hearts website. When Gavin learned of the affair he demanded to see the e-mails which she and the lawyer had exchanged and was appalled by what he saw—naked photos of each of them—and by what he read, which explicitly detailed what they would like to do to each other. Joanne said these were only fantasies and agreed to end the relationship, but it continued in secret. Her marriage was over in any case, and she resolved to leave Gavin and take the children with her. He planned otherwise.

In the early hours of 29 November 2005 Gavin took Milly downstairs, where he gave her an anti-depressant pill then cuddled her until she became drowsy. He then placed a cloth soaked in chloroform over her face and watched her die. He left the bodies of the family's two pet cats next to her. These he had killed with chloroform the previous day. Before he murdered Milly, Gavin had sent a text message to his wife, asleep upstairs, saying that he loved her and that Milly wished to remain with him. Once Milly had died he sent messages to his mother and to his wife's lover, James. Finally he sent a second text message to Joanne saying: "Goodbye. Milly sends her love. She died at 3.32 am. She did not suffer. My pills and chloroform saw to that. Myself I expect to suffer. Love till death do us part, I said, and this I what I meant." He then took the rest of his anti-depressant pills, cut his throat and wrists and passed out. Joanne discovered the full horror of what he had done when she came downstairs the following morning.

However, Gavin Hall survived his suicide attempt, and stood trial for murder in 2006 at Northampton Crown Court. His defence was that he could not remember anything that had taken place that night. But on Thursday, 2 November 2006, he was found guilty and sentenced to life imprisonment, to serve a minimum of 15 years.

6.7   The Priest and the Prostitute

Father Joseph Daly was 37 years old, and although his calling as a Roman Catholic priest demanded lifelong celibacy, for Daly the call of his sexual urges could not be denied. His church was in Runcorn, Cheshire, and he found a willing young lady in Manchester, the 27-year-old Winifred Markland, to respond to his needs. She was a seamstress but supplemented her income as a prostitute, and it was in this capacity that she hooked Father Daly in September 1876, when she travelled to Runcorn to ply her trade. He arranged to visit her lodgings in Manchester on Monday,

2 October at 10 pm. Winifred shared her rooms with a friend, Fanny Stevenson, who agreed to spend the night at her sister's home. As luck would have it, and despite the late hour at which he arrived, Daly was seen entering Winifred's house by someone who recognised him.

When Fanny returned the following morning she found the front door locked and noticed that Winifred was still in bed, apparently in a deep sleep because one arm was hanging over the side. Fanny thought it best not to disturb her, and took the same view when she returned home later in the day and found Winifred still in bed. Fanny went to visit a friend and stayed with her overnight. When she returned home the following morning she saw that Winifred had not moved and realised she was dead. Fanny called a doctor, who in turn called the police. A bottle by the bed was retained for analysis and the body taken to a mortuary for post-mortem examination. The bottle was found to contain nothing other than gin, which was also present in Winifred's stomach—but there it was accompanied by 48 grains (about 3 g) of chloroform.

The forensic analyst who presented this evidence in court was able to confirm that the chloroform had not resulted from her taking Chlorodyne, because he was unable to detect the other components of this medicine. It was concluded that someone must have added chloroform to her drink, and since there was no sign of chloroform in the house when it was searched it must have been brought in and removed afterwards. Daly was named as almost certainly the last person to see Winifred alive, and he was interviewed by the police. The coroner at the inquest allowed Father Daly to take the stand, where he declared his innocence and produced three independent witnesses who could vouch for his alibi. In his summing up the coroner declared that Winifred had been murdered, but not by Daly. There was no indication of suicide, because there was no sign of the bottle from which she might have taken the chloroform. Winifred Markland's murder was straightforward forensically. Clearly chloroform had killed her, and there was no doubt that it had been mixed with gin.

So was Daly guilty? Almost certainly. Had Daly's parishioners not perjured themselves to save him he might well have been tried for murder, found guilty, and hanged. He was already under suspicion of having a previous illicit affair. When he had first arrived in Runcorn he had seduced the 18-year-old daughter of a parishioner with whom he had stayed while looking for lodgings. After he had secured his own accommodation the girl continued to visit him and shared his bed. Together they planned to rob her father, who owned a grocer's shop, but they would make it look like an outside job. But the plan was discovered. The girl was thrown out on to the street by her father, and the priest was arrested and sent for trial, during which he was found guilty of stealing £12 and sentenced to 18 months in prison with hard labour.

6.8   The Murder of Edwin Bartlett

This case is more complex. It has continued to fascinate and divide crime writers as to whether the supposed perpetrator was guilty or not. The courts said not—but logic suggests otherwise.

Many believed Adelaide Bartlett to be the innocent victim of her husband Edwin's carelessness, and the jury agreed. The Pimlico Mystery, as the case became known, is the most famous murder case in which chloroform was responsible. What transpired in their rented rooms at 85 Claverton Street, Pimlico, London, on New Year's Eve, 1885 can only be surmised, but there was sufficient circumstantial evidence to convince the authorities that Edwin Bartlett had been murdered by his wife. She had the motive, the means and the opportunity. But the mystery about the method still remains: how did she manage to persuade her husband to drink chloroform? It was assumed that it was this that killed him. It had—but then again it hadn't.

Adelaide Bartlett was trapped in a marriage from which she longed to escape. Her husband, Edwin, was in the process of establishing a successful chain of grocery shops. However, she was spending a lot of time with a friend of her husband's, a Methodist minister, George Dyson, who was giving her lessons in classical subjects such as Greek and Latin. By the conventions of the time there was virtually no way she could escape from her marriage other than by the death of her husband, so she decided to poison him, at first unsuccessfully using **lead acetate**, and then successfully with chloroform.

Adelaide de la Tremoille was born in 1855 in Orleans, France. Her parents were Adolphe, a teacher of mathematics, and Clara, daughter of a clerk at the London Stock Exchange. (Her real father is thought to have been a French nobleman, the Comte de Thouars d'Escury.) When she was 19, Adelaide came to England to live with her uncle. He soon tired of taking responsibility for his niece, and found her lodgings with Charles Bartlett and his wife in Kingston-upon-Thames. Charles's 30-year-old brother, Edwin, fell in love with the beautiful Adelaide and they married in April 1875—against the advice of Edwin's father, who clearly saw what his besotted son could not, that Adelaide was self-centred and manipulative. Initially it was indeed a marriage in name only, as Adelaide went off to boarding school to complete her education and Edwin was occupied in building up his business. They eventually set up home together in 1877.

A clue about Adelaide's character was provided the year following, when she began an affair with Edwin's younger brother Frederick, or at least that is what Edwin's father concluded when it came to light that she and Frederick had been away for a few days, even though ostensibly not in each other's company. On her return she said she had been staying with an aunt, and husband Edwin was happy to believe this. Adelaide insisted that her father-in-law should not only apologise for his unjust accusations, but to sign a document confirming that these were completely untrue, and this he did. However, Frederick's subsequent emigration to America seemed to suggest that something had indeed been going on.

Maybe Edwin senior had hinted to his son that it was time Adelaide had a baby, and she did in fact become pregnant in 1881 but the baby was stillborn. Then in 1884 her wandering eye alighted on another young man, whom Edwin was cultivating as a friend: George Dyson, a young Methodist minister. Edwin enjoyed George's company and he became a frequent visitor, even agreeing to the suggestion that George should tutor Adelaide at his lodgings, where they then often had lunch together. Then in 1885 George was offered a chapel in Putney, at that time on the outskirts of London, but his association with the Bartletts continued when they took lodgings in the nearby district of Pimlico.

Sometime during the autumn of that year Adelaide resolved to be rid of her husband. She now told George—naturally in the strictest confidence—that Edwin was seriously ill and was likely to be dead within the year. The first part of this statement was a lie, but the second part was later to prove true. In November, Adelaide poisoned him using lead acetate. On account of its sweet taste this substance was known as sugar of lead. The doctors who carried out the post-mortem on Edwin noted a blue line around the gums, an acknowledged indication of lead poisoning, in fact it was often observed in those exposed to lead as a result of their employment, such as house painters. Adelaide may have heard of this poison as that used by Louise Taylor to kill Mary Ann Tregillis two years earlier in Plumstead, southeast London, for which she was executed in 1883.

In November 1885, Edwin suddenly experienced gut-wrenching stomach pains, another symptom of lead poisoning. He took to his bed and sent for a local doctor, Dr Alfred Leach, who noticed the blue line at the margin of Edwin's gums. Leach mistakenly thought this might indicate mercury poisoning and that Edwin was suffering from syphilis and treating himself with mercurial pills, the only effective treatment. Edwin strenuously denied both that he had syphilis and that he had ever taken mercury. Dr Leach's examination confirmed the absence of venereal disease and he was assured by Edwin that he enjoyed normal sexual relations with Adelaide. Lead acetate was a slow cumulative poison, more debilitating than deadly, and it required repeated doses over a prolonged period. It would be effective if the murderer was prepared to be patient, but Adelaide was not.

Edwin started to recover and Adelaide allowed his father to visit him, but he insisted on a second opinion. The new doctor came to the conclusion that Edwin was a hypochondriac, and advised him to leave his sick bed and get out of the house more. Dr Leach agreed with this advice and even suggested that the Bartletts take a holiday at the seaside—but that was the last thing Adelaide wanted. She wanted rid of her husband, and chloroform might be a more effective way of achieving this. During December Edwin's false teeth were causing trouble and Dr Leach persuaded him to have dental treatment, including removal of the roots of some of his remaining teeth. He was in considerable pain and could not sleep. Chloroform was an obvious remedy.

Mrs Doggett, the wife of the landlord, who lived in the downstairs flat, paid a social call on New Year's Eve. Adelaide asked what she knew of chloroform, but Mrs Doggett could reply only that she had once experienced its effects, nothing more. Adelaide was curious about chloroform and had persuaded George Dyson to purchase it, saying it was the only thing which seemed able to calm Edwin during the spasms caused by his supposedly terminal condition. Dyson believed her and obligingly bought 1 oz[iv] bottles at three different pharmacies, saying he needed it to remove stains from clothes; he combined them into one bottle, which he gave to Adelaide. The scene was now set for the murder. What transpired during the following eight hours has been the cause of much speculation.

Since his first illness Edwin had taken to sleeping in the parlour by the fire with Adelaide in a chair beside him, and it was there that he died. At 4 am on 1 January 1886 she roused the maid, Alice Fulcher, saying that she feared Edwin was dead and asked her to fetch Dr Leach. Adelaide also awoke the landlord, and he rushed upstairs to help. His observations are crucial. He found Edwin's body to be cold, suggesting that he must have been dead for several hours. He noted that the room smelled of chloroform, arising mainly from a brandy glass on the mantelpiece and from the traces of brandy around Edwin's mouth and on his chest. He also noticed that fresh coals had been put on the fire, and this was burning brightly.

When Dr Leach arrived he was told by Adelaide that she had wakened to find Edwin face down on the little bed, and had turned him over and poured brandy down his throat to revive him. Leach knew she was lying: Edwin's body temperature showed that he had died on his back, and a dead person is unable to swallow. At this stage these were only minor factors. When Edwin's father arrived later that morning he demanded a post-mortem and the body was carried into the back bedroom. Meanwhile Adelaide asked the maid to remove the brandy glass and wash it.

The following day, 2 January, a post-mortem was carried out by a Dr Green and observed by Dr Leach, and they noticed the strong smell of chloroform when the stomach was opened. The latter was removed for further examination. The two doctors now treated the death as suspicious and Adelaide was informed that she must immediately leave, because the apartment was being sealed until further investigations had been completed. Adelaide was then taken by George Dyson to stay with her friend, Alice Matthews.

George Dyson was soon aware of what had been found at the post-mortem and asked Adelaide about the bottle of chloroform,

---

[iv] One ounce (oz) is equivalent to 28 g.

but she told him to stop bothering her. The following day, Sunday, as he walked to chapel he threw away the three small bottles he had purchased. Again he went round to see Adelaide, and once more brought up the issue of the chloroform. A row ensued, overheard by Alice Matthews, in which Dyson reminded Adelaide that she had forecast Edwin's death, a fact which she now vehemently denied.

On 6 January Adelaide, George, and Alice Matthews went to Claverton Street to collect Adelaide's possessions. They then went to see Dr Leach, and while he and Adelaide were in conversation George learned the awful truth from Alice Matthews, that Edwin had not been fatally ill at all and that Adelaide had lied to him. The following day he again tackled her about the chloroform, but she replied that she had poured it away the previous day and thrown out the bottle.

In fact, a large part of the chloroform had ended up in Edwin's stomach. The analysis of its contents revealed the remains of his last meal, plus the alcohol he had taken, and the residue of at least an ounce of chloroform. Edwin had indeed consumed chloroform and brandy, but this had taken place some time before his death because some of the chloroform had passed into his intestines. The effects of this cocktail would have been little different from that of brandy on its own, although Adelaide's defence barrister contended that the mixture would have been so fiery that he would have realised that something was wrong and would have spat it out. If this had been the case then he must have deliberately drunk chloroform to commit suicide. From what we know of chloroform, drinking it is unlikely to kill a person so quickly. Despite what the court was told, Adelaide would probably have had little difficulty in getting her husband to take the chloroform, because it does not burn the throat when mixed with other components. Even today there are over the counter medicines which contain it.

The Bartletts had in their home a well-used copy of *Squire's Companion to the British Pharmacopoeia*. This was produced during the inquest into Edwin's death and was seen to fall open automatically at the page for chloroform. It was indicated there that chloroform evaporated readily and left no unpleasant odour, and that it could be dissolved in brandy, in which form it could be used as a sedative, although the recommended dose was as little as 5 minims (0.3 ml).

Adelaide probably persuaded Edwin to consume chloroform with brandy and may even have shown him the bottle that his friend George had procured for that very purpose. We know that the Bartletts had Chlorodyne in the house, and what Adelaide was offering could simply be described as a stronger version of this well-known cure. Maybe Edwin thought that drinking a little might kill the pain he was suffering, but in the event it had little effect. Maybe Adelaide then suggested that he should breathe the vapour. What Edwin no doubt died of was *breathing* chloroform, and it is not difficult to imagine him agreeing to this if it would help him sleep. All it needed was a handkerchief dipped in chloroform over his face, and within five minutes he would be asleep. A little more chloroform and he would enter second phase anaesthesia, a little more and he would be snoring deeply, and finally, more chloroform would ensure sleep from which he would never awake. One can imagine Adelaide watching as he stopped breathing and then leaving the chloroform-soaked handkerchief over his face long enough to be sure there was no sign of life. Perhaps she even fell asleep herself. That would explain why Edwin's body felt cold when the Doggetts saw him at 4 am. Edwin probably died sometime before midnight.

Adelaide now had to act the part of the innocent. The smell of chloroform within the room would be strong, so maybe she opened the window for a while, or until the room felt cold. She threw the handkerchief on the fire and added more coals. The fact that the fire was newly stoked with coal was commented on by those she called. Even so, she had not completely eliminated the smell of chloroform, and that too was noted. Before anyone came she had to show that she had attempted to save her husband, and so she splashed brandy round his face, even succeeding to drip some into his mouth. Her story was simple: she had fallen asleep and awakened to find her husband dead. Some time in the night he had taken chloroform in brandy to sooth his pain, but he had taken too much. By 1885 it was well known that chloroform could be dangerous and he had killed himself with it, either accidentally or intentionally.

Adelaide had concocted what she thought was a reasonable explanation for her behaviour during the previous weeks. Her friendship with George Dyson was of course platonic. Nevertheless, at Adelaide's trial the maid said she had more than once found them in situations not normally associated with tutor and pupil. On one occasion she had observed Adelaide fondling George's head in her lap as he sat at her feet on the floor.[v] Adelaide even claimed that Edwin had said that if anything happened to him she should marry George. In any event Edwin had been a far from adequate husband, in fact she claimed she and Edwin had only once had sex in their married life. By 'sex' Adelaide meant *procreational* sex; others took it to have its more normal meaning of *recreational* sex, which is what Adelaide's defence wanted them to think.

In fact Edwin and Adelaide had led a normal sexual life, and what she was indicating was that they only once had *unprotected* sex, which was when she conceived. Of course in the atmosphere of Victorian Britain it was unthinkable for anyone to be questioned about their sex life, and her version was designed to suggest that Edwin's inadequacy had depressed him to the point at which he had become suicidal. Adelaide's implication that they led a celibate life was to be denied by Annie Walker, the midwife who attended Adelaide. What the defence hoped would not be revealed in court was that condoms were found in one of Edwin's pockets.

In Victorian times condoms were available, but judging by the large size of many families—six or more children were quite common—they were either unreliable or not often used. Condoms had been discreetly on sale in England from the 1600s, when they were originally made from a region of a sheep's intestines which had a closed end. The rubber condom appeared in the mid-1800s and it was these which were found among Edwin's possessions. Condoms could be purchased at shops which sold medical items, since technically they were in use to prevent the spread of disease. There was no mention of them in the popular literature and information was spread only by word of mouth. Indeed, the word condom was considered too embarrassing even to appear in the Oxford English Dictionary until 1972. The subject must have been infinitely more embarrassing in 1885, but the revelation that condoms had been found rapidly undid the impression created by her counsel that their marriage was one in name only.

---

[v] The Reverend Dyson even composed a short poem to Adelaide which suggested he had fallen in love with her:

Who is it that hath burst the door,
Unclosed the heart that shut before,
And set her queen-like on its throne,
And made its homage all her own – my Birdie.

That myth, carefully fostered by Adelaide's counsel, Edward Clarke QC, was exposed as nonsense when the nurse who had attended her during her labour was called a second time to the witness box. Her cross-examination proceeded as follows:

Clarke: At the time you nursed Mrs Bartlett in her confinement, did you become aware from anything she said to you with regard to its having been the result of a single [sexual] act?

Nurse: Yes, sir.

Clarke: Thank you, that will be all.

The judge, Mr Justice Wills, then asked the nurse what it was that Adelaide had actually said.

Nurse: That it happened only once—on a Sunday afternoon.

Judge: She said so?

Nurse: Both of them said that there was always some preventative used.

Judge: You say you heard that from them both?

Nurse: Both of them.

During the judge's summing up he reminded the jury of the testimony of one of the detectives who said that he had found four 'French letters'—a euphemism for condoms—in Edwin's trouser pocket. The judge dismissed the idea that the Bartlett's marriage was a marriage in name only, but Clarke's closing speech for the defence was a masterpiece of courtroom oratory and it prevailed. No jury could condemn a beautiful woman like Adelaide to the hangman's rope and they found her not guilty on the evidence as given. (They expressed doubts as to whether she really was innocent.) The result was that Adelaide was acquitted. She disappeared into obscurity, taking her husband's wealth with her. Dyson had served his purpose, and was dumped.

CHAPTER 7

Life & Death & CO: Carbon Monoxide and the Home-Made Gas Chamber

A word shown in **bold** *indicates that further information will be found in the Glossary*

7.1 The Silent Killer

Carbon monoxide (CO) is a gas which kills more people than any other poison, not as an agent of murder but more often by accident. Breathing air containing 0.1% carbon monoxide can within half an hour convert 50% of the haemoglobin in the blood to the useless carboxyhaemoglobin, which destroys its capacity to carry oxygen from the lungs to the muscles and the brain. Remain in such an environment for a further half an hour and you will almost certainly die. Carbon monoxide has been described as the silent killer because it gives no warning of its presence, nor are its effects recognised by victims, who can pass from a state of slight nausea and a headache to unconsciousness within the space of a few minutes. Even if death is avoided, brain damage may still be extensive.

How carbon monoxide does this is well established, but it has come as a surprise in recent years to discover that the human body actually *needs* a small amount of carbon monoxide, in fact it makes its own. There is every likelihood that CO-releasing drugs will one day feature in medical treatment, delivering carbon monoxide to

Molecules of Murder: Criminal Molecules and Classic Cases
By John Emsley
© John Emsley, 2008

reduce inflammation or increase the flow of blood by relaxing the artery walls. In the case of organ transplants it has an important role in preventing rejection by suppressing the immune system.

Deadly though carbon monoxide can be, few murderers have chosen it, but there is one tragic tale and a second rather ingenious story to tell. The tragic tale concerns a Los Angeles man who murdered five of his six children, and the ingenious one concerns a magistrate in England who tried to gas his wife and make it look like suicide.

7.2   The Properties and Uses of Carbon Monoxide

Carbon monoxide is a colourless, odourless, poisonous gas. The French chemist Joseph de Lassone (1717–1788) was the first to produce it in a pure condition, and did so in 1776 by heating a powdered mixture of zinc oxide and coke (carbon). He noted that it burned with a blue flame, which he assumed indicated hydrogen, a gas which had only recently been discovered, and he took his discovery no further. It was left to an English chemist, William Cruickshank (1745–1810), in 1800 to identify the gas as a new oxide of carbon.

Carbon monoxide is not just a deadly gas, it is an important industrial chemical and has been essential for the chemical industry for more than a century. Carbon monoxide can be made from carbon or from methane (CH_4) by reacting them with oxygen or steam. Coke was the source of carbon when coal was the fossil resource of the chemical industry. Blowing oxygen through a bed of glowing coke at around 800 °C will generate carbon monoxide provided there is not too much oxygen,[i] and the process is self-sustaining. Blowing steam in place of oxygen will also produce carbon monoxide along with hydrogen gas,[ii] and the resulting mixture is known as synthesis gas, or **syngas**. While some carbon monoxide is still made from carbon, more is likely to be made this way as natural gas supplies dwindle. Sometime in the future carbon monoxide will have to be made from renewable sources of carbon such as charcoal derived from biomass or from methane derived

---

[i] The reaction proceeds in two stages: $C + O_2 \rightarrow CO_2$, followed by $CO_2 + C \rightarrow 2CO$.
[ii] The reactions are $C + H_2O \rightarrow H_2 + CO$, and $CH_4 + H_2O \rightarrow 3H_2 + CO$.

from anaerobic oxidation. Meanwhile, natural gas will continue to be used to make carbon monoxide because 50 trillion cubic metres, of the 170 trillion cubic metres of known natural gas reserves, are in regions too remote to be piped to centres of population, although some is shipped as liquefied natural gas. It makes more sense to convert the methane to syngas and turn this into higher value products and ship those, and such chemical conversion plants are in operation in places like Alaska, Trinidad, and Malaysia.

The carbon monoxide in syngas can be converted to hydro-carbon fuels, methanol or **carbonyl chloride**, all of which are important industrially. Plants to convert it to transportation fuels have been operating since the 1920s. It is most easily converted to methanol, which could be blended with petrol to give a cleaner burning motor fuel, but since methanol is toxic there is limited enthusiasm for this. However, when methanol reacts with further carbon monoxide in the presence of a catalyst, it forms acetic acid (or **ethanoic acid**, as it is now termed), which is used in the manufacture of a wide range of consumer products, including paints, **silicone** resins, **polyester** fibre, and **poly(methyl methacrylate)** plastics. Seven million tonnes of acetic acid are produced this way annually.

As a chemical, carbon monoxide is simple in structure but complex in application. It is as simple structurally as nitrogen gas (N_2), in that it consists of only two atoms, has the same number of electrons, and has a triple bond between the two atoms. The two gases are similar in boiling point: liquid carbon monoxide boils at $-192\,°C$ and liquid nitrogen at $-196\,°C$. Complexity results from the asymmetry of carbon monoxide, which makes it a polar molecule. The non-bonding pair of electrons on the carbon atom can become attached to metal atoms, giving carbonyl compounds, for instance with nickel it gives nickel carbonyl, $Ni(CO)_4$. This compound is formed whenever nickel makes contact with carbon monoxide, and special care has to be taken when handling the gas to be sure this does not happen. On the other hand, this unwanted interaction has a beneficial outcome, giving rise to a new route for the production of nickel and ultimately to the manufacture of better, and cheaper, stainless steel.

The story begins in 1888 when Ludwig Mond (1839–1909), a leading chemical manufacturer, found that nickel valves, which he assumed would not corrode, were being attacked by an impurity in the carbon dioxide used in one of his processes. The impurity was carbon monoxide, and he set one of assistants, Carl Langer, the task of finding out what was going on. He reacted finely powdered nickel with carbon monoxide in a sealed tube, and when this was opened he vented off the carbon monoxide and ignited it. He was surprised to see that the gas burned with a bright green flame, and when the flame made contact with a porcelain dish a film of pure nickel was deposited. Almost unbelievably, the nickel had been converted to a gas—in fact, nickel carbonyl is a liquid, boiling at 43 °C, but its volatility contradicted existing chemical theory, which was that no metal compound could behave as a gas. Investigation revealed that nickel and carbon monoxide react at 50 °C to form nickel carbonyl, and that if this is heated to 180 °C it breaks up into its constituent parts.

The upshot of this was the development of the Mond process for refining pure nickel, purer than it had ever been made previously. The growing demand for stainless steel, which contains as much as 35% nickel,[iii] meant that there was a ready market. By 1900 the Mond Nickel Company was in profitable production in South Wales, but there was a heavy price to pay in human terms due to an occasional accidental leakage of nickel carbonyl vapour. Over the years a number of workers were affected, and some died—but not of carbon monoxide poisoning, as analysis of their blood showed only slightly raised levels of this gas. What killed them was the film of nickel which formed in the lungs and caused the generation of fluid. As this filled the lungs, breathing became increasingly difficult and eventually ceased altogether. Tests on animals showed that nickel carbonyl was five times more deadly than carbon monoxide, albeit requiring more time to achieve the same end.

Carbon monoxide has other uses, one of which is for preserving the colour of meat, especially when packaged, so that cuts of beef and pork retain a fresh appearance. The carbon monoxide reacts with the myoglobin in the meat to form a red pigment which is more resistant to oxidation and retains its colour. This technique was approved by the **FDA** in 2004, on the grounds that the improved colour did not act as a preservative. On the other hand some consumers are unhappy with its use in the belief that it might mask the odour of meat which is going off. The carbon monoxide

---

[iii] Plus varying amounts of chromium.

itself poses no threat to a person eating the meat. And, as we shall see, the human body is itself no stranger to this gas.

7.3 Carbon Monoxide in Nature

Carbon monoxide has always been a natural component of the Earth's atmosphere because it comes from volcanoes, some of whose escaping gas can contain as much as 2%. Carbon monoxide naturally makes up between 0.05 and 0.2 ppm of the atmosphere and there are about 500 million tons of the gas circling the globe. It comes from bacteria, forest fires, and the oxidation of hydro-carbon molecules like methane. The atmosphere also has a carbon monoxide component as a result of human activity and this comes from vehicle exhausts, heating boilers, and industrial smoke and adds 450 million tons to the atmosphere each year. Where all the carbon monoxide ends up is still not completely resolved, but most is oxidised to CO_2 and is absorbed by the leaves of growing plants or by the water of the sea as part of the larger carbon cycle of Nature. Soil also appears to be a natural 'sink' for carbon monoxide, and soil microorganisms are thought to absorb large amounts although how and why they do this has yet to be understood.

Almost everything that burns produces some carbon monoxide and that is true of tobacco smoke as well. Roughly half comes from the burning of fossil fuels, and half from the burning of wood, straw, and the clearing of forests. The largest human contribution is car exhausts and this can result in levels as high as 50 ppm (0.005%) in very heavy traffic. None of this carbon monoxide poses a long term threat because the gas is fairly rapidly oxidised to CO_2, and its residence time in the atmosphere is about two months.

7.4 Physiological Effects

The poisonous nature of the fumes given off from red-hot charcoal was known as long as 300 years ago and was mentioned by Hoffmann in his book, *Considerations of the Fatal Effects of the Vapour from Burning Charcoal*, published in 1716. A scientific study of the toxicity of carbon monoxide was carried out by the Frenchman, Claude Bernard, in the 1840s. He used it on dogs, then dissected them to study its effect. He immediately noticed that their blood was bright scarlet. As we have seen, carbon monoxide forms a bond with haemoglobin to form carboxyhaemoglobin, which has a much brighter colour than the normal oxygenated form. Low-level carbon monoxide poisoning produces a headache and flu-like symptoms, higher levels give dizziness, and dangerous levels lead to unconsciousness, as the central nervous system is affected, and, once the heart muscle is no longer supplied with oxygen, death soon follows. Even if an unconscious victim is quickly removed to safety there may be permanent damage to a number of organs, including the brain.

Exposure to small but constant amounts of carbon monoxide over a period of time produces a variety of symptoms, which may at first not be recognised as having been caused by this gas. In addition to headaches and muscular pain there can be depression, chronic fatigue syndrome, and even hallucinations. In some instances this has been used to explain experiences associated with haunted houses, such as feelings of dread, sudden shivering, strange noises resembling footsteps, and even some unexplained deaths. In certain cases the haunting has been traced to a defective boiler or heater which is emitting carbon monoxide. Once the fault has been corrected these phantoms of the night are generally exorcised. The Victorian era was a golden age for haunting and weird manifestations, and one wonders to what extent this was a consequence of the widespread use of town gas for lighting and cooking. This was piped to most homes and buildings and it inevitably contained a few per cent of carbon monoxide. The replacement of town gas by natural gas might serve to explain why the spirit world no longer appears to be so active, although the idea of a spirit world still haunts some people.

Haemoglobin is the body chemical most affected by carbon monoxide because the CO molecule readily combines with the iron atom in the centre of haemoglobin—in fact haemoglobin binds almost 250 times more strongly to carbon monoxide than it does to the oxygen molecule, O_2. The advantage of haemoglobin as an oxygen carrier is that it is readily able to release the oxygen, whereas it clings strongly to carbon monoxide. In fact, there are four oxygen-binding sites within haemoglobin, and if just one of these picks up a carbon monoxide molecule the ability of the other three sites to transport oxygen is reduced. Carbon monoxide has the strongest effect on those organs which are most dependent on a consistent supply of oxygen, the brain and the heart. Even after a person appears to be recovering from carbon monoxide poisoning, the damage to these organs may not be evident for several days. Brain damage may be apparent as short term memory loss, irritability, slurring of speech, Parkinson-like symptoms and even dementia. These delayed symptoms are more likely in the elderly and are more pronounced if the person was rendered unconscious due to the effect of carbon monoxide.

How a person responds to carbon monoxide depends on factors such as the rate of breathing, whether they are anaemic, and the presence of pre-existing heart disease. Survival is mainly determined by a person's ability to remove themselves from the danger. This may not be possible if they are disorientated and their movements are uncoordinated. In this case, breathing air containing as little at 0.05% of the gas can be fatal.

At concentrations in the air below 35 ppm no detectable symptoms are likely. Concentrations between 35 and 200 ppm can cause a headache within a couple of hours, increasing in intensity as the level of carbon monoxide increases. At 800 ppm a throbbing headache occurs within half an hour, followed by dizziness and nausea, and unless action is quickly taken, unconsciousness will occur within an hour. At 1500 ppm death is likely within two hours, at 3000 ppm within half an hour, and at 6000 ppm a person feels faint in a minute or two and is dead within 20 minutes. Finally, at 12 000 ppm (1.2%) it needs only a few breaths to become unconscious and the heart will cease to beat within ten minutes.

How much carbon monoxide a body has absorbed can be measured by analysing the proportion of carboxyhaemoglobin present in the red blood cells. In normal health there can be as much as 5% of haemoglobin in this form and for heavy smokers the figure can rise to 10%. Even if 10% of haemoglobin is in this form there are still no observable symptoms. At 20% a person finds breathing more difficult and feels slightly light-headed. At 30% there will also be a headache and at 50% the person begins to stagger and their movements are uncoordinated; above this level they are likely to pass out, and finally at 60% they will probably die.

Haemoglobin is not the only natural chemical in the body which attracts carbon monoxide—myoglobin is also deactivated by it. This substance stores oxygen in the muscles so that during strenuous exercise, when the energy demand cannot be met from oxygen in the blood, this reserve supply is available. The *cytochrome oxidase* enzyme present in the mitochondria is also rendered inactive by carbon monoxide, affecting the oxidation processes and energy formation within the cell. The attraction of carbon monoxide for this enzyme is less strong, however, and the cell can switch to an alternative way of generating energy and it goes into anaerobic mode – at least for a time.

The brain regulates breathing according to the concentration of carbon dioxide in the blood; it does not directly monitor oxygen levels, nor decide how much is needed. For this reason a person unknowingly breathing carbon monoxide may not notice anything wrong, and they continue breathing normally until they suddenly lose consciousness. If they die, it is immediately obvious at autopsy that they have been affected by carbon monoxide because the colour of the skin is a healthy pink (assuming they are white) and the blood will be the bright red of carboxyhaemoglobin.

After a person has been rescued it normally takes about four hours for the level of carbon monoxide in the blood to fall by 50%, but this can be cut to one hour if pure oxygen is administered. When treating a person who has been overcome by carbon monoxide this is the first step to take, ideally in a pressure chamber at three atmospheres. The half-life of carboxyhaemoglobin is around 5 hours, but breathing pure oxygen can reduce this to about 1.5 hours; in a pressure chamber the figure comes down to 30 minutes.

Accidental deaths from carbon monoxide come from badly ventilated gas fires, furnaces, heaters, wood-burning stoves, motor vehicle exhausts or camping stoves. Carbon monoxide can occur whenever organic material burns in a deficiency of oxygen. Prevention is better than cure, of course, and this explains why all gas-burning appliances must now be checked annually by a qualified engineer. Nevertheless, a carbon monoxide detector is also advisable for rooms where such appliances are used. A detector can measure either a build-up of the gas, or a slow accumulation over time. In some American states it is mandatory to have such devices installed, as in Ontario, Canada, and in New York City. The first carbon monoxide detectors were unpopular in inner cities as there was a higher background level in the atmosphere, and this was sufficient to trigger the detector after a few weeks with as little as 15 ppm in the air. Alarms are now only activated if the level reaches 70 ppm, or 30 ppm over a period of 30 days.

The simplest detectors are visual devices in which palladium chloride reacts with carbon monoxide to form a dark-coloured complex. More sophisticated detectors use a solid state detector comprising tin oxide (SnO_2), which is very sensitive to carbon monoxide and its conductivity changes, triggering the alarm. Another type of detector uses an electrolytic cell. If the electrolyte absorbs carbon monoxide the conductance changes in proportion to the quantity present. These detectors can be used to measure accurately the level of carbon monoxide in a sample of air.

7.5 Positive Features of Carbon Monoxide

Thomas Beddoes (1760–1808) founded the oddly named Pneumatic Institution for Inhalation Gas Therapy in Bristol, England, in 1798 and treated a range of ailments, including tuberculosis, paralysis and venereal diseases, by getting patients to breathe air containing a variety of vapours and gases. One of the gases was carbon monoxide, which made people appear healthier by giving them rosy cheeks. This was later regarded as a risky experiment, indeed downright foolish in view of the known toxicity of carbon monoxide. However, Beddoes could have been on to something, since carbon monoxide has the ability to lower blood pressure and dilate the blood vessels, and it is possible that one day it might be used for this purpose if a suitable form of medication can be found which releases carbon monoxide when and where it is needed.

It had long been accepted that smokers were less likely to suffer from high blood pressure, and some wondered whether this might be caused by the carbon monoxide present in tobacco smoke. However, other factors could have explained the observation. For example, smokers are less likely to be overweight and therefore to have high blood pressure, but the nicotine present tends to produce the opposite effect. What the carbon monoxide appeared to be doing was counteracting the effect of nicotine. It was, however, difficult to separate out the effect of carbon monoxide from the other factors. In 1991 a paper from the Department of Physiology at Wayne State University School of Medicine, Detroit, Michigan, discussed the issue in detail, pointing out that some researchers detected a link between carbon monoxide and lower blood pressure, whereas others found the two to be unrelated. Indeed, some researchers even reported that carbon monoxide *increased* blood pressure. The paper from Wayne State came to the conclusion that if carbon monoxide had any effect on blood pressure it would be to decrease it. We now know they were right.

That carbon monoxide occurs naturally in the human body has been known for a long time. That it might actually have a *positive* role to play seems barely credible, but there is little doubt that this is true. The carbon monoxide in the body comes from the haemoglobin itself. If no more than 15% of the haemoglobin is carrying carbon monoxide, this actually benefits the body by reducing inflammation, widening blood vessels and thereby increasing blood flow, and lowering the clotting tendency of blood. It is also known that carbon monoxide is an important factor in organ transplantation, since it suppresses the activity of macrophages, which are a feature of the body's rejection mechanism.

Somewhat unexpectedly, the carbon atom in carbon monoxide produced naturally by the body originates from a CH grouping in the haemoglobin molecule itself. This group can be oxidised to carbon monoxide with the assistance of the enzyme, *haeme oxygenase*—in fact all mammalian cells constantly generate carbon monoxide in this fashion.[iv] It is now thought that the carbon monoxide may play the role of neurotransmitter. Some bacteria also generate carbon monoxide, but in this case from carbon dioxide, and they possess a specific enzyme to achieve this named *carbon monoxide dehydrogenase*, which contains both iron and nickel atoms at its active site. The purpose of bacterially generated carbon monoxide is unclear, but again it could form part of a signalling pathway. *Carbon monoxide dehydrogenase* is found in very primitive bacteria and might well have been in existence when significant amounts of carbon monoxide were present in the Earth's early atmosphere, when the conditions were anaerobic. This gas is known to be present in space, indeed analysis of the tail of the Hale–Bopp comet confirmed its presence.[v]

The theory that carbon monoxide might be able to prevent the onset of certain inflammatory conditions was first aired in 1998. Carbon monoxide has been given to patients in low doses before transplant surgery, since it gives protection against organ rejection.

[iv] When exposed to the solvent dichloromethane (CH_2Cl_2, also known as methylene chloride), the body metabolises its $-CH_2-$ component to CO.
[v] I am indebted to Professor Brian Mann for this information.

In balloon angioplasty it is seen to reduce blockage of the pulmonary arteries. In 2006 researchers at the Beth Israel Deaconess Medical Center, part of the Harvard Medical School, discovered that carbon monoxide was able to relieve pulmonary arterial hypertension (PAH), a debilitating condition eventually causing heart failure. The pulmonary arteries carry blood from the heart to the lungs, and normally their walls are smooth. If the walls thicken and hinder the flow of blood a blockage eventually occurs, resulting in a heart attack. Until recently PAH was thought to be incurable, but the Harvard group speculated that carbon monoxide might be able to alleviate it, thanks to its known ability to provide relief for high blood pressure caused by other arterial conditions. Accordingly they tested it on a mouse with PAH, and exposed the animal to carbon monoxide for one hour daily. Not only did this relieve the animal's condition—it actually cured it.

For some reason carbon monoxide not only controls the overproduction of cells in the artery walls, it can kill the rogue cells while leaving the normal cells unaffected. The researchers discovered that it was able to achieve this selectivity with the help of another two-atom molecule, nitric oxide (NO). Further research, using arterial cells grown on an artificial membrane, showed that the two types of cell were somehow in communication, and that the healthy cells were generating nitric oxide and initiating the process known as programmed cell death in the proliferating cells. The question of *why* this was happening remains unanswered, but this has not hampered small-scale trials being conducted on humans.

Carbon monoxide is now regarded as an endogenous gasotransmitter, a name given to small gaseous molecules produced by the body with specific roles. These are nitric oxide (NO), hydrogen sulfide (H_2S), and now carbon monoxide (CO). They are all highly toxic when inhaled in air. Carbon monoxide causes relaxation of the smooth muscle cells of arteries, either by opening the channels through which potassium ions can flow, or by acting on enzymes such as *cytochrome P450*. It also suppresses **endothelin**, which reduces blood pressure. Collaboration between researchers at Sheffield University, led by Brian Mann, and at Northwick Park Research Institute in London, led by Roberto Motterlini, has resulted in methods of delivering carbon monoxide to the specific points in the body where it is needed. In this way it may in due course be possible to prevent the formation of blood clots which trigger heart attacks, and even reduce the inflammation which causes arthritis.

Mann has discovered water soluble molecules which release controlled amounts of carbon monoxide when injected or taken orally. These carbon monoxide-releasing molecules (CORMS) are based on ruthenium, iron or manganese. Most metal carbonyls are toxic, soluble only in organic solvents, and are willing only release carbon monoxide during chemical reactions. What was needed was a compound which was non-toxic, soluble in water, and able to give up its carbon monoxide as required. Tricarbonylchloride (glycinato)ruthenium(II), known as CORM-3, fulfilled these three requirements. In experiments with living cells and on live rats, it was shown to act in conjunction with other natural chemicals in the body to reduce blood pressure. It releases carbon monoxide in a controlled manner and prolongs the survival time of transplanted organs, reducing the risk of rejection.

Other carbonyls have been designed to release carbon monoxide over a period between 30 minutes and several hours, depending on the condition being treated. A company, hemoCORM, has been set up to commercialise the new drugs.

7.6 The Deadly Danger of Carbon Monoxide

In the 1800s and up to around 1960, suicides often took their lives by breathing in carbon monoxide. It was customary for a would-be suicide to put their head in the gas oven and turn on the taps. In a few minutes they would be unconscious and within half an hour they would be dead. During this period gas was also the cause of innumerable accidental deaths as families went to bed in a house with a leaking gas pipe, which was sometimes to be found in bedrooms where gas was used for lighting. They would be discovered dead by friends and neighbours the following day. Town gas, which was made by heating coal, contained as much as 8% carbon monoxide. When town gas was replaced by natural gas then this kind of accident became impossible, but other accidents with carbon monoxide still occurred because incomplete combustion of natural gas can generate it.

Some years ago I came across the sad tale of a family that had been inadvertently killed by a blackbird. In early spring 1993, Michael and Deborah Mason set off from London with their two young sons, Christopher (4) and Jeremy (2), to spend a weekend at a cottage they owned in north Devon. The time of year was significant, because a pair of blackbirds had built a nest in the flue of the gas boiler in the cottage and the female was already sitting on a clutch of eggs. Her choice of nesting site had the unfortunate effect of blocking the ventilation system. That weekend Michael, Deborah, Christopher and Jeremy all died of carbon monoxide poisoning. Deaths from carbon monoxide have declined markedly in the UK, falling over the past 11 years from 74 in 1996 to only 17 in 2007, thanks to rigorous testing of the ventilation of gas-burning appliances and the introduction of carbon monoxide detectors.

This kind of tragedy is repeated every year around the world. The greatest risk of carbon monoxide poisoning arises in new accommodation or in poorly maintained holiday apartments, when it is not appreciated that previous occupants may have sealed the ventilators, or that a flue or chimney has become blocked. When natural gas is denied sufficient oxygen it burns with a yellow flame and if used for cooking leaves sooty marks on the sides of pans. In the USA around 10 000 people a year are affected by carbon monoxide from badly ventilated gas appliances and need hospital treatment, and of that number around 250 die. Four times as many as this deliberately kill themselves with this gas, mostly sitting in the garage in their car with the engine running. Exhaust fumes from modern cars no longer offer quite the same threat of accidental poisoning, since a good catalytic converter can reduce the carbon monoxide content of the exhaust gases to below 0.25%, whereas at one time they could account for as much as 20%. Nevertheless, in an enclosed space such as a garage with the door closed, the available oxygen is eventually used up, and even a catalytic converter cannot prevent an increasing amount of carbon monoxide being produced by the engine.

The worst accidental mass poisoning from carbon monoxide was on 2 March 1944 in Balvano, Italy, when a packed train stalled in the Armi Tunnel and 521 people died. The train was also pulling freight cars and the engine was burning low-grade coal, and as it approached the tunnel the wheels began to slip and the train came to rest with the passenger cars inside the tunnel and the last couple of freight cars out in the open. The build-up of carbon monoxide overcame the engine driver, and then slowly filled the confined space of the tunnel until everyone on the packed train had succumbed. There have been other examples of multiple carbon monoxide poisonings, although not on the scale of the Balvano incident. One of these took place in December 1995 at the Hill of Calvary Missionary Baptist Church in Detroit, Michigan, when 23 worshippers were overcome by fumes from a faulty boiler. Ten years later, in March 2005, a similar accident occurred in the same city, this time affecting 48 worshippers at the multi-faith Sai Center, where the service was being held in a room rented from a Catholic day school. Thankfully their supplications were heeded, and in neither of these incidents did anyone die.

A number of famous people have met their death from carbon monoxide, including the American novelist and poet Sylvia Plath (1932–1963), whose struggle with depression was integral to her writings. Her marriage to the English poet Ted Hughes was not a success; they separated after he had an affair and she moved with her children to London. On the morning of 11 February 1963 she sealed the kitchen with wet cloths around the door in order to protect her two sleeping children, and then put her head in the gas oven.[vi] There is some evidence that Plath did not intend to kill herself and that her suicide may have been more a cry for help, because she left a note for a Mr Thomas in a neighbouring apartment asking him to call a doctor, but he was unaware of the note and took no action. The French novelist, 62-year-old Emile Zola (1840–1902), was also killed by carbon monoxide. He died on 29 September 1902 as the result of a blocked chimney and his death was so unexpected and suspicious that some at the time felt he had been murdered. (He left no suicide note.) It was not until many years later that a man who had been repairing the roof admitted that he had deliberately closed a vent on the chimney—because he passionately disagreed with Zola's political views.

In 1947, Britain's most famous forensic pathologist, Sir Bernard Spilsbury, committed suicide using town gas in his laboratory in University College, London. By then he was 70 years old and was suffering the after-effects of a series of strokes, and he had lost both of his sons during World War II. It was a sad end to an illustrious career.

Cassandra was the daughter of King Priam of Troy and a prophetess who had been given the power to foresee the future by the

---

[vi] This was before the time when the UK converted to natural gas.

god Apollo. But she was cursed by him after resisting his amorous advances, the curse being that in future no one would believe anything she said. She foretold the defeat of Troy, but her warnings went unheeded. Those who advocate the fitting of carbon monoxide detector alarms in every home must feel rather like modern-day Cassandras. Even if these alarms were made compulsory by law, it would not deter those intent on murder from removing or disabling them before they took action. In the next section of this chapter we will take a look at two carbon monoxide crimes, both of which oddly involved people who had been named after the ancient prophetess. Like her, they survived the disaster that was planned.

7.7   The Murder of the Garcia Children

On the evening of 19 February 2002, Adair Javier Garcia, a 30-year-old technician with a telephone company, put his five children to bed in the bungalow where they had lived for only two weeks in the Whittier area of Poco Rivera, a town of 65000 inhabitants located ten miles southeast of Los Angeles. The children had had a lovely day, and before they went to bed he asked them to tape messages for their mother, who had recently decided that family life was not for her and had left. In the tapes they described how they been to see the movie, *Peter Pan 2: Return to Neverland*, and had had a meal at the ChuckE.Cheese fast food restaurant on Rosemead Boulevard. They did not realise it would be their last message to their mother.

Had the movie so affected their father that he believed that he and his children might fly away to a wonderland together? In any event, on the way home he bought a bag of charcoal and some matches. After the children had filmed their messages he put them to bed, closed all the windows and tacked blankets over the outside doors. He then lit the charcoal in a barbecue grill and placed it in the passageway outside the bedrooms. He recorded his own farewell message to his wife on video:

> "I love you. I am of sound mind and I know what I am doing. I just want to let you know that this is the only option available. You've broken me completely. You don't care, so I don't care. What I'm about to commit is the most cowardly, selfish act possible."

He then went to bed. That night Brenda (aged 10), Jonathan (7), Vanessa (6), Cecilia (4), and Anthony (2) succumbed to carbon monoxide poisoning. The family members were discovered the following morning by Garcia's mother-in-law, Adriana Arreola, when she arrived to take the older children to school. She found the house full of smoke and at first had difficulty finding the children. She assumed there had been a tragic accident, and simply opened windows to let out the smoke, but she quickly realised she should phone the emergency services. When the sheriff's deputy arrived at the house he found three of the children still breathing, but two of them died shortly afterwards. Garcia himself and just one of the children, Kassandra (9), survived and were taken to the nearby Whittier Hospital. Adriana called her daughter, the children's mother, who came quickly to the bungalow and collapsed in grief at the tragedy that confronted her. It quickly became clear to the police that this was no accident.

After he had been discharged from hospital Garcia was arrested and brought to trial in March 2005 before Judge John Torribio. The trial lasted only a few days, and a tape was played to the jury in which Garcia admitted intending to kill himself and his children as a way of getting back at his wife, who had ended their marriage only a few days previously. The evidence given by the officer first on the scene so affected one of the women jurors that she was unable to continue, and had to be replaced. The jury of seven women and five men deliberated for seven hours before finding Garcia guilty of the murder of the five children who had died. The tape the children had made was not shown during the trial itself but was viewed by the jury when it came to deciding the penalty to be imposed. Although his was a capital offence, the jury recommended that Garcia be spared the death penalty and on 8 June 2005 he was sentenced to life imprisonment.

What was surprising about the Garcia murders was that he and Kassandra survived, which suggests that the level of carbon monoxide inside the bungalow that fateful night probably was not much above 500 ppm (0.05%).

7.8 Margaret Jones's Nightmare

Mr Cranog Jones was 44 years old and had originally trained as an engineer. He had been on a number of management courses and was now head of education and training at Smiths Industries, a defence company based in Cheltenham, England. He married his wife, Margaret, in August 1972 and they had three children, one of whom, Cassandra, still lived with her parents in the tiny, one-pub village of Brockhampton, a few miles east of Cheltenham. Mr Jones was a respected member of the community and in 1986 became local magistrate for the Cheltenham district court. In his spare time he was a do-it-yourself enthusiast.

But all was not as it appeared, and in the autumn of 1991 Mr Jones had only recently returned to the family home having been living for some time with his mistress. (Nor was this his first extra-marital affair. It seems that something similar had happened during a postgraduate course in management studies at Bristol Polytechnic in the 1990s.) In December 1991 he began to apply his DIY skills to devising what was intended to be the perfect murder of his wife. Despite his leaving her several months previously, he had become obsessed with wanting to prove that she too had been guilty of an adulterous affair and had even hired a private detective to follow her movements and tap her phone to confirm his suspicions. But nothing came of these investigations. Consequently any divorce would have been to his disadvantage, both financially and in terms of the occupancy of their home.

On the night of Saturday, 7 December 1991, Jones's plan was ready to be put into action. He had drilled a large hole through the outside wall of his wife's bedroom, ostensibly to install a sink and vanity unit, and through this hole he fitted a plastic pipe leading over a conservatory and down the wall at the back of the garage. Although Mrs Jones was unaware of it a second hole had been made through the garage wall so that the pipe could be connected to the exhaust of her car, a Ford Fiesta, inside the garage. That night when his wife had gone to bed Jones put his plan into operation. He crept into the bedroom where Margaret was asleep and quietly connected a two-foot square box to the pipe leading through the bedroom wall. He then went down to the garage and started the car engine, with the choke at its maximum to ensure a fuel-rich mixture in order to generate a high proportion of carbon monoxide.

All that remained was to return to the bedroom, arrange the box over his sleeping wife's head, wait for her to become deeply unconscious, and carry her body to the car. He would then disconnect the hose from the exhaust, remove the tubing, leave the car engine running and close the garage door. Within an hour or so she would be dead, and all the problems of getting a divorce and the monetary details that tend to complicate such events would now be avoided. A few days beforehand he had tried out his device on his daughter Cassandra's cat, and the animal was dead within minutes.

His plan went wrong as soon as he went upstairs to the bedroom. He realised too late that he should have muffled the sound of the exhaust, since the box was making a loud hissing noise, sufficient to waken his wife from her deep sleep. She found her husband crouching at the side of the bed. Realising what he was trying to do, she immediately leapt out of bed and ran to phone the police, only to find that the wire had been pulled out of its socket and the phone was dead. She ran from the house to a public telephone along the road and called for help. She dared not return home, and crouched down behind the wall of a chapel and waited. Jones, realising that his plot had failed, dismantled the pipes and the box, plugged the hole in the bedroom wall, and put his murderous contraption in the boot of his car. He checked that all appeared normal and drove away, dumping the portable gas chamber and pipes on the local refuse tip.

When the police later questioned him Jones denied everything that his wife had told them, but later went back to the police station to admit that he had in fact dumped the tubing and piping. He now claimed that he had used it merely to vent exhaust fumes from the garage while he was working on the car.

Jones eventually divorced his wife in July 1992 and was free to marry his 28-year-old lover, Diane, who also worked at Smith Industries, and they set up home in Cheltenham. At the time, Jones was still on police bail having already been charged with attempting to murder his wife.

The trial took place in April 1993 before Judge Starforth Hill QC. The prosecution was in the hands of Christopher Clark QC and Jones was defended by Michael Parroy QC. Mrs Jones told the court that on the night in question she quickly realised what her husband was trying to do and accused him of wanting to kill their daughter, Cassandra, as well. In court Jones admitted lying to the police about disposing of the piping and other apparatus at a local rubbish tip, but mentioned in mitigation that he had voluntarily admitted this to the police later.

In his final address to the jury the defence counsel related Jones's story, that the hole in the bedroom wall was for an outlet pipe for a vanity unit and that Mrs Jones's version of events was mere fantasy. Indeed, if he had intended to gas her as she claimed, why did he need to drill a hole through the bedroom wall when all he needed was to insert the pipe through the bedroom window? Nor was she the nice person she appeared to be in court—he said that she had another side to her character and that she had deliberately framed him so that she could have the family house for herself. His explanation of what he was doing in the garage late that night was that he had had a row with his wife, and became so resentful of all the time he had put into mending her car that he decided to undo what he had done; he had been in the garage no more than five minutes when he heard his wife shouting from the bedroom window. As an explanation it did not quite ring true.

On Thursday, 23 April, the jury at Winchester Crown Court deliberated for six hours before finding him guilty. Judge Hill sentenced him to nine years in jail, saying: "The method you used to try to kill your wife was ingenious, required preparation, and had been well thought out. It was to be a death without any sign of physical injury, capable of giving the appearance of suicide." He also reminded Jones that he had betrayed the trust implicit in his position as a magistrate, namely to uphold the law.

There will no doubt be other murderers in the future who try to kill their victims with carbon monoxide, but as an agent of homicide this has little to recommend it as both Adair Garcia and Cranog Jones found out.

CHAPTER 8

Cyanide and the Death on the Nile

A word shown in **bold** *indicates that further information will be found in the Glossary*

8.1 The Speediest of Ends

Every month the chemical industry manufactures enough hydrogen cyanide (or prussic acid, as it was once known) to kill every person on the planet. Thankfully, in practice hardly anyone dies as a result of this abundance of toxic material—indeed it is even used to make pharmaceuticals which can save lives. The few who are murdered with cyanide, or those who poison themselves with it, are dead within minutes, as it is the quickest acting of all poisons. There are antidotes, but these have to be given immediately. Nature herself also manufactures cyanide, and this has caused harm to a lot of people in the past, especially as the result of eating cassava, a starchy root crop, when it has not been properly prepared. This is a staple food in many tropical countries.

Cyanide has been used to commit crime, and we will examine the case of the Birkenhead solicitor, Cheryl Lewis, who met her death in a hotel room overlooking the Nile at Luxor. We will also review the most famous—and unsolved—cyanide poisoning carried out in recent years, in which the murderer added it to bottles of the popular painkiller Tylenol in the USA, and in so doing not only killed a number of innocent individuals but also produced an upheaval in the packaging industry, and indirectly suggested a way

Molecules of Murder: Criminal Molecules and Classic Cases
By John Emsley
© John Emsley, 2008

in which other would-be murderers might be able to murder and get away with it.

Deadly though cyanide usually is, some have survived it, none more remarkably than the mystic monk, Rasputin. He was given cyanide and survived. Was it his spirituality that saved him, or could there have been a chemical explanation?

8.2   The Chemistry of Cyanide

The **cyanide** structure involves the combination of a carbon atom and a nitrogen atom by means of a triple bond, which comprises three electron pairs depicted as $C{\equiv}N$. This grouping is strong, but highly reactive. It is able to form stable organic compounds known as nitriles, for example acetonitrile, CH_3CN. Nitriles have nowhere near the toxicity of inorganic cyanides. The latter contain the negatively charged cyanide ion, CN^-. Salts such as sodium cyanide (NaCN) and potassium cyanide (KCN) are soluble in water, and in solution are extremely toxic. Even more toxic is the corresponding free acid, hydrogen cyanide (HCN). This is a volatile colourless liquid carrying a faint odour of bitter almonds and has a boiling point of only 26 °C, which means that on a warm day it is a gas. Dissolved in water it is weakly acidic, which accounts for its older name, prussic acid. Hydrogen cyanide is the deadliest of all the cyanides, and traces can be fatal if breathed in. Prussic acid was named by the Swedish chemist, Carl Wilhelm Scheele (1742–86), who first produced it in 1782 by heating the pigment **Prussian Blue** with sulfuric acid.

Prussian Blue is itself perfectly safe and has been in use for centuries, despite being a cyanide of iron.[i] This pigment gave cyanide its name, which comes from the Greek *kuanos*, meaning dark blue. Somewhat unexpectedly, Prussian Blue has turned out to be the only effective antidote for thallium poisoning. It has also been used to treat those exposed to radioactive caesium–137, which is released when nuclear power plants go critical, such as that at Chernobyl in the USSR in 1986.

---

[i]Prussian Blue is nominally ferric ferrocyanide, $Fe_4[Fe(CN)_6]_3$, as are Chinese Blue, Bronze Blue and Milori Blue. The analogous compound ferrous ferricyanide, $Fe_3[Fe(CN_6)]_2$, is the pigment Turnbull's Blue. I am grateful for this information to Don Sanders, formerly Chief Colourist, Ciba–Geigy Pigments.

Cyanides are natural chemicals. Indeed a little is essential to the workings of the human body because it is a component of vitamin B_{12}, and of the *[NiFe]-hydrogenase* enzyme which catalyses the oxidation of hydrogen. Vitamin B_{12} has at its centre a cobalt atom to which the cyanide residue is bonded. Although the daily requirement for vitamin B_{12} is only 2 µg, which is a minute amount, it is vital to health as its absence leads to pernicious anaemia; it also has other important roles in the body, including the synthesis of the nucleic acids in DNA and of the amino acids in proteins. Some vitamin B_{12} is derived from bacteria in the intestine and the rest comes from food such as milk, beef, eggs and salmon.

The cyanide ion forms strong bonds with metals, due in part to its negative charge, and also to its ability to form so-called π-bonds.[ii] A particularly important ionic material is gold cyanide, $[Au(CN)_2]^-$, which is one of the very few compounds formed by this metal and is the reason why cyanides are used in the extraction of gold from its ores. If the ore is high grade it is stirred in large vats with a solution of sodium cyanide, whereas low grade ores are sprayed with this solution and the liquor collected and concentrated. The gold is recovered by adding zinc dust, or by adsorption on to activated carbon. Although large amounts of cyanide are required in this process there is little threat to the environment since it is all recycled. Any which may accidentally escape into the atmosphere as hydrogen cyanide decomposes under the effect of oxygen and sunlight into carbon dioxide, nitrogen and water.

Hydrogen cyanide may have had a part to play in the creation of life when the Earth was still young. In the 1950s Stanley Miller and Harold Urey carried out an experiment to see whether the building blocks of life could have formed naturally from the simple molecules present in the Earth's early atmosphere. They subjected a mixture of methane, ammonia, hydrogen and water vapour to electrical discharges simulating lightning—and amino acids were created. The route by which they were produced almost certainly involved the formation of hydrogen cyanide, which then reacted to give more complex molecules.

Hydrogen cyanide, sodium cyanide and potassium cyanide are important industrially. About 1.4 million tonnes of hydrogen cyanide are produced annually, of which the USA accounts for half.

[ii] Typical examples exhibiting this type of bonding are the ions $[Fe(CN)_6]^{3-}$ and $[Pt(CN)_4]^{2-}$.

(Details of cyanide production are given in the Glossary.) It is used in the manufacture of a wide range of products, including cosmetics, fire retardants, paints and dyestuffs. About a third of the hydrogen cyanide produced annually goes into **adiponitrile**, which is used in the manufacture of the polymer, **nylon**. Another polymer derived from hydrogen cyanide is **poly(methylmethacrylate)**, examples of which include the transparent plastics known as Perspex and Plexiglas. About 15% of hydrogen cyanide output is converted to sodium cyanide for the mining industry. A lot of hydrogen cyanide goes into the manufacture of the amino acid, **methionine**, which is used to enrich animal and poultry feeds.

Generally speaking, once the cyanide group forms a bond with a carbon atom in a larger molecule, as in an organic nitrile, it loses its toxicity. This structure occurs in a number of pharmaceuticals, including the anti-ulcer drug **cimetidine** (Tagamet), the angina drug **verapamil** (Isoptin), and the antidepressant **citalopram** (Celexa). These drugs are non-toxic due to the cyanide radical being covalently rather than ionically linked. On the other hand some organic nitriles do exhibit toxicity, for example the insecticide **cypermethrin** (sold as Ripcord or Barricade), which is deadly to insects but not to animals. In rare instances the cyanide radical bonded to a metal can even be a life-saver. **Sodium nitroprusside** is one such compound,[iii] and is occasionally used in emergency medical treatments and in medical research to produce a rapid decrease in blood pressure by dilating the blood vessels.

8.3   Cyanides in the Diet

Cyanides are produced by bacteria, fungi, algae, insects and plants, including those such as almonds and spinach which form part of the human diet. Apple seeds contain cyanides, and 100 g of crushed apple seeds contain 200 mg of the substance **amygdalin**, a glycoside which takes its name from *amugdale*, the Greek for almond, which contains a cyanide group in its molecular structure. In theory it would be possible to break down this quantity of amygdalin to release around 10 mg of hydrogen cyanide. Other fruits which have seeds containing traces of cyanides include cherries, apricots, peaches and almonds. Some millipedes release hydrogen cyanide as

---

[iii] Sodium nitroprusside has the chemical formula $Na_2[Fe(CN)_5NO]$.

a defence mechanism, and certain insects such as day-flying moths also contain a deadly dose of cyanide to deter predators. Cassava root contains **linamarin**, another natural molecule containing both glucose and cyanide groupings, and again the cyanide content may be present in order to protect the plant—but this does not prevent it being a staple food for humans.

Because our early primate ancestors ate a mainly vegetarian diet containing a variety of glycosides, the enzyme *rhodanese* evolved which rapidly detoxifies the cyanide radical by converting it to a thiocyanate group (–SCN), which is a thousand times less toxic. (The sulfur atom comes from the amino acid glutathione.) *Rhodanese* is present in the mitochondria of every human cell, enabling a healthy adult to detoxify as much as 1 g of cyanide over a period of 24 hours.

Cassava is popular because it grows in poor soil and is pest and drought resistant. It is a key crop in many regions of the world and is eaten as a staple food by as many as 500 million people, some consuming as much as 0.5 kg a day as a kind of porridge. This is made by peeling and grating the root, soaking it in water for three days or more, drying it in the sun, and finally grinding it into a flour which can be used to make the porridge. This food is extremely popular in Africa, but too much of it can cause a disease called *konzo*, cases of which first appeared in the 1930s and became more common in the 1980s, especially in regions afflicted by drought.

Symptoms of *konzo* include paralysis of the legs, difficulty in speaking, dizziness, headaches and vomiting. These sometimes disappear within a few days, but in some cases people remain permanently disabled. At one time doctors were puzzled by this and thought they were dealing with an infectious disease, since it only affected specific households and villages. Eventually it was concluded that *konzo* was due to chronic cyanide poisoning caused by eating too much cassava. In theory there is enough cyanide present in a single cassava root to kill a number of people, so why does it not poison everyone who eats it? It depends on the way the food is prepared.

The linamarin in cassava root is broken down in two stages. Grating the root releases an enzyme from the plant cells which catalyses the hydrolysis of the linamarin into glucose, **propanone** and hydrogen cyanide; the hydrogen cyanide either slowly evaporates or is leached out into the water used for soaking, which is then discarded. Fresh cassava root contains 300 mg of cyanide per kg, but if it is peeled and soaked in water for a day and then sun-dried for four days, this falls to 80 mg per kg. If peeled and soaked for five days and then sun-dried for a week, the figure becomes 30 mg per kg, and grating before soaking brings it down to 10 mg per kg. The World Health Organisation recommends that the last of these figures should be regarded as the maximum allowable limit for cyanide in cassava flour. This would ensure that a person eating 500 g of cassava root would receive less than one-tenth of the lethal dose of cyanide, estimated to be around 50 mg.

The extent of a person's exposure to cyanide can be assessed by determination of the amount of thiocyanate ion (SCN^-) in the urine, this being the metabolic product which the body forms in the liver from cyanide. Analysis of the urine from those in a village affected by *konzo* has shown thiocyanate levels in the range 25–60 mg per litre, compared to the 3–6 mg per litre expected for those in an unaffected village.

Konzo outbreaks have occurred when food has become too scarce to give time for the shredded cassava to be soaked sufficiently, or where a variety of bitter-tasting cassava with a higher level of linamarin has been used. Cyanide poisoning cannot provide the complete answer, however, since in some instances women are affected by *konzo* but their menfolk are not, even though both have a similar diet. Villages near the coast were not usually affected. Fish in the diet appeared to offer some protection which it might have done by providing protein rich in the sulfur-containing amino acids methionine and cysteine. These make it possible for the body to convert cyanide to thiocyanate and thereby aid its excretion.

8.4   Toxicity of Cyanide, and its Antidotes

The lethal dose of cyanide for an adult is considered to be around 50 mg, but cases have been known in which doses as high as 250 mg have not proved fatal. Cyanide is toxic due to its ability to bind to the iron atom present in the active site of the enzyme, *cytochrome oxidase*, which occurs in the membrane of cell mitochondria. This enzyme is essential to life since it catalyses the final stage of the oxidation of glucose by oxygen, vital for releasing its energy. If the enzyme is rendered inactive by cyanide, the final transfer of electrons from the enzyme to oxygen is blocked and the cascade of chemical reactions that provide the energy for vital processes cannot proceed, with the result that life quickly ebbs away. The central nervous system and the heart are immediately affected when this source of energy dries up. The oxygen transported from the lungs by the haemoglobin in the blood cannot then be used, which explains why one of the symptoms of cyanide poisoning is the presence of bright red oxygenated blood in the veins. Cyanide is an irreversible enzyme inhibitor, which means that once it has bound itself to the iron atom in the enzyme it can only be removed by chemical attack. The victim becomes unconscious but remains breathing, although this slowly ceases as a deep coma develops. The pulse becomes so weak that it can hardly be discerned, and finally the heart stops altogether. Large doses of cyanide can be fatal within minutes, although some victims have been known to take several hours to die—and, as we shall see, in rare instances they may even survive.

Plants have a different *oxidase* enzyme in their mitochondria, which may be less efficient than the animal version—but it is immune to cyanide. As a result, plants are insensitive to concentrations of cyanide which would kill animals, and a few plant species are even known which contain lethal quantities of cyanide, such as the shoots of the giant bamboo. However, the golden bamboo lemur can eat these without coming to any harm, even though they contain enough cyanide to kill most other animals.

When it comes to toxicity, hydrogen cyanide is in a class of its own. The maximum allowable concentration of hydrogen cyanide in air is 10 ppm. Breathing air with 150 ppm for 30 minutes can endanger life, and at 300 ppm death is virtually certain. Half of the population of rats exposed to 170 ppm hydrogen cyanide were dead within 30 minutes, and at three times this level (510 ppm) all had died after 5 minutes. On 8 February 1924 the first execution using hydrogen cyanide was carried out, on Gee Jon in the State of Nevada. He had been a hit-man working for one of the Chinese tongs. The gas was generated inside the sealed execution cell by dropping pellets of potassium cyanide on to a tray of concentrated sulfuric acid beneath the chair in which the condemned man was strapped. Fumes of hydrogen cyanide were liberated, and death occurred within minutes.

Hydrogen cyanide can be produced as the result of combustion. It is present in tiny quantities in tobacco smoke and vehicle exhaust fumes, but it can also be formed in dangerous concentrations when acrylic polymers[iv] burn; these are present in many household fabrics. The smoke from house fires can cause hydrogen cyanide poisoning.

Early in the last century, when it was necessary to rid premises such as houses, barns, greenhouses or the holds of ships, of vermin such as rats, cockroaches, bedbugs or fleas, all the entrances were sealed and hydrogen cyanide was pumped in. Inevitably this led to accidents, especially when the premises were later re-entered and some hydrogen cyanide was still present, for example within absorbent items such as mattresses. Remarkable instances of survival have however been recorded, including one in 1938 in which a man appeared to be dead after contact with the hydrogen cyanide he had been spraying inside a building. He was given artificial respiration for 90 minutes, after which he appeared to be breathing again. He was given injections of sodium nitrite and sodium thiosulfate, and three days later he recovered consciousness. When used together, these two chemicals can act as an antidote to cyanide poisoning.

Even more dramatic was the story of a maintenance engineer working at the American Cyanamid Company in New Orleans, at which hydrogen cyanide was produced for acrylonitrile manufacture. He was disconnecting a pipe which normally carried hydrogen cyanide and which was supposed to have been purged, but still contained the liquid. He was drenched with it, to the extent that as it evaporated it actually caused his clothes to freeze. His workmates rushed him to the medical room and administered amyl nitrite directly into his nostrils. This substance dilates the blood vessels, maintaining the flow of blood to the heart muscle and the brain. The doctor in charge continued the treatment but the patient appeared to have died. Nevertheless, he suddenly recovered and seemed none the worse for his experience, even turning up for work the following day.[v]

An antidote has to provide the cyanide ion with an alternative iron-containing molecule to latch on to, so that it leaves the *cyctochrome oxidase* unaffected. The sacrificial molecule chosen is a

---

[iv] Acrylic polymers are based on poly(acrylonitrile), more correctly known as poly (propenonitrile).

[v] This remarkable story was told to me by Michael Utidjian, Corporate Medical Director of the company.

form of haemoglobin. The body contains sufficient surplus haemoglobin to mop up around 1000 mg of cyanide. Before the cyanide ion can combine with it, the iron atom must be in the Fe(III) state rather than its normal Fe(II) condition, and this is achieved by the antidote sodium nitrite. In this form it is known as methaemoglobin,[vi] and as such it becomes the iron to which the cyanide group prefers to attach itself. The sodium thiosulfate present then converts the cyanide component to the harmless thiocyanate ion (SCN^-), which the body can readily excrete in the urine.

Alternative methods for treating cyanide intoxication are preferred in some countries. For example, in France **hydroxycobalamine** (also known as pro-vitamin B_{12}) is employed, which is able to bind rapidly to the cyanide group. The latter converts it to cyanocobalamine (vitamin B_{12}), rendering it safe and in a form which can be excreted. Hydroxycobalamine is effective both within the intravascular space and within cells. This treatment is so effective that it is administered routinely to all victims of smoke inhalation, and is generally considered superior to alternative treatments for cyanide poisoning. An additional benefit of this antidote is that hydroxycobalamine has no effect on a person who has not been exposed to cyanide, whereas antidotes such as sodium nitrite can in these situations cause a catastrophic drop in blood pressure.

In Germany a different antidote, **4-dimethylaminophenol** (4-DMAP), is preferred, which acts more rapidly and has a lower residual toxicity. Intravenous injection of 3 mg per kg of 4-DMAP converts one-third of the body's haemoglobin to methaemoglobin within one minute.

Kelocyanor is another antidote, and this uses a different approach, employing cobalt to pick up the cyanide. It acts more quickly than methaemoglobin, but Kelocyanor has toxicity of its own and this drawback has led to accidents in the UK. Patients thought to have been poisoned by cyanide were given Kelocyanor, but when cyanide turned out not to be involved they suffered from cobalt poisoning instead. Nevertheless, Kelocyanor has a long shelf life compared to other antidotes and is therefore recommended by the UK Health and Safety Executive.

[vi] Too much of this in the blood is not desirable because this form of haemoglobin cannot transport oxygen.

Death from cyanide poisoning is unfortunately so rapid that medical help usually arrives too late. Anyone handling hydrogen cyanide is therefore advised to keep a capsule of amyl nitrite on hand in case of accident. This, together with artificial respiration, may keep a person alive long enough for the antidote to be administered. An indication that an unconscious person has been affected by cyanide is likely to be the smell of almonds on their breath.

It has also been suggested that sugar, or its glucose component, might protect a person against cyanide poisoning by reacting with it to form less toxic compounds which can be eliminated by the body. Indeed, a research study found that cyanide toxicity in mice was reduced when the cyanide was mixed with glucose. This might explain the apparent immunity of Grigory Rasputin (1872–1916) to the cyanide which his killers put in the sweet pastries and Madeira wine which he ate. Other attempts had been made to assassinate this powerful but malign influence on the Russian Imperial family, but none had succeeded—in fact this self-styled monk appeared to lead a charmed life. A group of aristocrats plotted to kill Rasputin on the evening of 29 December[vii] 1916 and they invited him to the Moika Palace, the home of Prince Felix Yusupov, where he was given the poisoned food. But nothing happened. The Prince then drew a revolver and shot him, but not fatally, and Rasputin escaped into the courtyard where he was shot again by another conspirator, Vladimir Purishkevich, an extreme right-wing member of the Duma, the Russian parliament. Still he did not die, so the two men tied him up, dragged him to the frozen Neva River and pushed him through a hole in the ice, where Rasputin finally drowned.

Did he really survive the dose of cyanide? One theory was that his stomach did not contain the usual digestive acids needed to convert the cyanide to the deadly hydrogen cyanide. Alternatively, he might have been saved by other ingredients in the food which he consumed. Or had the conspirators no idea how much cyanide constituted a fatal dose and used too little cyanide to achieve their aim? Or had the potassium cyanide they used already decomposed? This can happen when it is exposed to the atmosphere. Given time, water and carbon dioxide are able to convert it to potassium hydrogen carbonate ($KHCO_3$), enabling the cyanide to escape as hydrogen cyanide vapour and thereby rendering ineffective their supposed deadly poison.

[vii] This was December 16 in Russia, which had not yet converted to the Gregorian calendar.

8.5 Cyanide as a Non-Accidental Killer

Hydrogen cyanide has potential as a chemical warfare agent. It has the advantage that it can be stored indefinitely in the form of solid sodium cyanide and the poison gas can be generated as required. Sodium cyanide was stockpiled both by the USSR and the USA in the 1950s and 1960s. However, as a military weapon hydrogen cyanide is far from ideal since it is lighter than air and disperses quickly, making it difficult to maintain a fatal concentration over a large area. It is also easy to achieve protection by wearing a gas mask. Nevertheless it is listed as a chemical warfare agent under the Chemical Weapons Convention, and manufacturing plants in countries which are signatories, and which have a capacity greater than 30 tonnes per year, must be declared and remain open to inspection.

Hydrogen cyanide was the active agent in Zyklon B, first manufactured by the German chemical industry during World War I. It was a powerful fumigating agent and particularly effective for delousing bedding and clothing, and was imported into the USA and used for this purpose in the 1930s. It was supplied in sealed canisters containing liquid hydrogen cyanide absorbed in an inert material such as cardboard or the mineral anhydrite (calcium sulfate), together with an agent giving it a foul smell which ensured that those using it were aware of its presence. Zyklon B was also supplied as a fumigant to concentration camps in the Third Reich—but in 1942 it began to be supplied, minus the smell, for use in the gas chambers of the extermination camps in which many millions were slaughtered.

Suicide using cyanide has also claimed many lives, in one famous instance involving hundreds of women and children. This was the fate chosen in 1978 by the inhabitants of the Peoples' Temple in Guyana, when almost a thousand of its members willingly drank a fruit-flavoured solution of sodium cyanide. The village known as Jonestown had been set up by the cult leader, Jim Jones, as a self-sufficient religious community. In 1978, the village leaders felt threatened when a Democrat congressman, Leo Ryan, along with various other US officials and journalists, arrived to investigate complaints that human rights were being abused and that the community was run on highly authoritarian lines, including inhumane punishments. When they arrived at an airstrip near the colony the unwelcome visitors were attacked and the congressman and four journalists were killed. The following day, on 18 November 1978, the cult members assembled and were handed cups of the poison. A total of 913 drank it, after they had used it to murder their children, of whom 276 died.

Sodium or potassium cyanide is the perfect agent for those seeking a fast-acting suicide pill, and relies on the generation of hydrogen cyanide by the acid in the stomach. A glass capsule of hydrogen cyanide which can be crushed in the mouth is even quicker, and was the device preferred by some members of the Nazi hierarchy, including Heinrich Himmler and Hermann Goering. Even Adolf Hitler himself crushed a capsule in his mouth when he shot himself, having just witnessed his new wife Eva Braun commit suicide in a similar manner. Cyanide capsules have also been employed more recently. In 1987 Kim Sung Il bit into a cyanide capsule hidden in a cigarette after he had left a bomb on board a Korean airliner which exploded over the Indian Ocean on 29 November. Members of the Sri Lankan terrorist organisation, the Tamil Tigers, are also reputed to wear a capsule of hydrogen cyanide around the neck, which allows them to commit suicide if they fall into the hands of the security forces.

Other more worthy individuals have used cyanide to take their own lives, indeed two of the most distressing deaths were those of the scientific geniuses Wallace Carothers in April 1937, and Alan Turing in 1954. Carothers was a chemist employed by the US chemical company, DuPont. He was the first to produce nylon and demonstrated that it could be spun into fibres which were much stronger than natural fibres. Carothers however suffered from bouts of intense depression and it was during one of these that he locked himself in a hotel room in Philadelphia and drank a solution of cyanide. Alan Turing was the genius who solved the supposedly unbreakable German Enigma codes used in World War II. He used a primitive computer to do this and later worked in 1948 at the University of Manchester on their new computer, which at the time had a larger memory than any other computer in the world. However, he was arrested for homosexual activities, then a criminal offence in the UK, and was faced with the choice of prison or medical treatment. He chose the latter, but found it so depressing that he used cyanide to commit suicide at his home.

Novelists have also chosen cyanide as the agent of death. Australian author Nevil Shute's novel, *On the Beach*, is about life after a devastating nuclear war. In this the Australian government hands out cyanide tablets to the population so that they can kill themselves rather than face slow death from the cloud of radioactive dust which has obliterated the rest of humanity, and is now drifting their way. On a less depressing note, in the James Bond novels we discover that Agent 007, in common with other 00 agents, is issued with a cyanide capsule for use in the event of capture. Come what may, suicide was never going to be an option for James Bond and he throws his away. Murder by cyanide was also involved in Agatha Christie's detective novels, *And Then There Were None*, and *Sparkling Cyanide*. Raymond Chandler's gangster novel, *The Big Sleep*, also uses cyanide as the instrument of murder. This was turned into a supposedly classic *film noir* in 1946 starring the legendary Humphrey Bogart, although the director of the film clearly lost the plot on the cutting room floor. Real murders are equally fascinating, and slightly less confusing, and we will look at two in which sodium cyanide was involved.

8.6 The Murder of Cheryl Lewis

In 1974 John Frederick Allan obtained his degree in industrial chemistry at Hatfield Polytechnic, now the University of Hertfordshire. While at college he had fallen in love with a fellow student, Jacqueline Millard, and after graduation they married and emigrated to Zambia, where Allan obtained a post at a mine near the town of Chingola. The marriage was not to last, however, and they divorced in 1977, but what Jacqueline later recalled most clearly about the years in Zambia was her husband's hobby of collecting butterflies. His method of killing them was to put them in a large jar with a piece of moistened sodium cyanide. That way they died quickly without fluttering about and damaging their fragile and beautiful wings. Allan had no difficulty in obtaining sodium cyanide, which was readily available in the laboratory at work. He brought some pieces back with him to the UK—no doubt he thought it might come in useful one day. That day came 20 years later, in 1998, when he took some on holiday as part of a plan to solve some pressing financial problems.

Soon after the breakdown of his first marriage Allan fell in love again, this time with an Iranian woman named Sima, and they married. One night, Sima was mysteriously shot and wounded, and Allan was arrested. However she assured the Zambian police that her husband had not been responsible and he was freed on bail, whereupon he returned to the hospital, carried his wife from her bed to his car, and with their bags already packed they left for South Africa and thence flew to the UK. They set up a small printing business in Birkenhead, a city across the Mersey from Liverpool, settled down and raised a family. However, family life was not satisfying enough for Allan.

At the age of 40 he began an affair with Cheryl Lesley Lewis. She had recently divorced her husband and was making a successful career as a solicitor in Birkenhead. In 1993 Allan obtained a divorce from Sima and moved in with Cheryl at her home, The Poplars, Woodbank Park, in a leafy suburb of Birkenhead. (They never married.) She was a senior partner in the firm of solicitors known as Cheryl Lewis and Company, of Westbourne Road, which employed a number of staff. As befitting her wealth and status she drove a large Mercedes with the personalised number plate, A9 CLL. Allan told her friends that he was a businessman and that he was writing two books, one a guide to golf courses and the other about corruption in South Africa. What he didn't reveal was that his main source of income was now from social security. None of his writing enterprises made any money, and he relied on Cheryl to pay the bills. Occasionally they visited Cheryl's parents at Pwllheli in north Wales, who later admitted they did not like her new partner, but said nothing at the time because they did not want to alienate her. They were perfectly right not to trust Allan, because his intentions were purely dishonourable, and their fears were confirmed when they learned on Friday, 9 October 1998, that their daughter had suddenly died—in Egypt.

Allan and Cheryl had gone there for a holiday and were staying at the New Winter Palace Hotel near Luxor.[viii] That evening, their last in Egypt, they enjoyed a meal in the restaurant overlooking the Nile and then went to their room, No 508. While Cheryl was in the shower Allan poured her a gin and tonic to which he added cyanide,

---

[viii] This is a five star hotel on the banks of the Nile and so named to distinguish it from the old Winter Palace, built in 1886 when Egypt was still part of the British Empire.

and within minutes of drinking it she was writhing in agony. Allan went down to the hotel's reception desk in an apparently distressed state and said he needed a doctor very urgently to attend to his wife. It just so happened that an American tourist, 28-year-old Pamela Black, was on hand; she had been trained in first aid during the time she worked for an airline, and she offered to help.

They rushed up to Room 508 and found Cheryl barely conscious, lying completely naked on the bed, sweating profusely, with foam coming from her mouth and with a curious blue pallor to her skin. When Pamela felt Cheryl's pulse it was very faint and an odd odour appeared to be coming from her mouth, and indeed seemed to permeate the whole room. Pamela could see that Cheryl was in a serious condition and asked Allan to give her mouth-to-mouth resuscitation. This he (wisely) declined to do, saying he preferred to wait for the doctor to arrive, which happened 15 minutes later. Pamela later commented that Allan had seemed unwilling to do anything for his wife, even to the extent of not helping to put some clothes on her. The doctor who came realised that Cheryl was beyond help and called an ambulance to take her to hospital. On arrival she was pronounced dead. The death certificate indicated the cause of death as heart failure and acute coronary thrombosis. Cheryl's body was later embalmed and flown back to the UK.

A representative from the holiday travel company also visited Room 508 and she too noticed the blueness of Cheryl's skin, which she said was visible beneath her suntan. Allan told her that he thought a notorious Liverpool drug dealer called Curtis Warren might have been responsible for Cheryl's condition. She had come across this man three years previously when she had allowed the police to install surveillance equipment in her office to observe the comings and goings at his premises across the street. (Warren was later jailed in the Netherlands for drug dealing.)

Cheryl's family were naturally stunned by what had happened and insisted on a full autopsy, since the supposed cause of death was entirely at odds with their 43-year-old daughter's normal state of health. The autopsy was carried out by pathologist Edmund Tapp, but he was unable to confirm the reasons given on her death certificate, indeed he found no indication that Cheryl had been suffering from any disease that might have killed her. What he did find, however, was that her stomach was inflamed as if by some corrosive agent. On the other hand there was no similar inflammation in her mouth and gullet. These findings were consistent with her having swallowed a drink laced with sodium cyanide, which would only have formed the corrosive prussic acid in contact with the natural acid in her stomach. The pathologist's opinion was that Cheryl had been poisoned, possibly with cyanide, but he could find no direct evidence of cyanide in her body. Cyanide disappears from the body tissue after death, but in any case she had already been embalmed, removing all traces of poison from her stomach. Moreover, Cheryl was a heavy smoker, which meant that small amounts of cyanide found in her body could possibly have originated from cigarette smoke. Allan knew full well that it would be virtually impossible to prove that Cheryl had been poisoned by cyanide.

The police were now involved because it was beginning to look like a case of murder. When they interviewed Allan he rather boastfully said: "I'm an applied chemist and chemical engineer. I could have given her something and I'd know what to give her so that it wouldn't be found." Others were later to recall that he had made a similar boast when speaking about the comedy, *The Wimbledon Poisoner*, first shown on BBC television in 1994, about a man who manages accidentally to kill lots of friends and acquaintances while trying to poison his overbearing wife, in this case with thallium. Allan commented that he knew a much better poison than thallium which would do the job and leave no trace. Although he did not specify what it might be, he was clearly referring to cyanide.

On his return to Birkenhead Allan continued to live in Cheryl's house and drive her car. He also set in motion the legal procedures that would allow him to inherit her wealth according to her will. Cheryl's parents had been unaware that she had written a new will and, suspecting it might not be genuine, they informed the police. This information now provided them with a motive, increasing their suspicion that the murderer was Allan. He appeared to be acting in ways that confirmed this—for example, he reported that Cheryl's Mercedes had been stolen. When this was traced by the police they discovered in one of its compartments a plastic bag containing two gold rings and two gold bracelets, along with documents relating to their valuation, plus four small lumps of sodium cyanide, three of which appeared to have had small pieces knocked off them, and the fourth had been deliberately cut and a portion estimated to be about 10 g removed. This was of course the sodium cyanide that Allan had brought back with him from Africa many years earlier. He maintained that the car had been stolen and the plastic bag must have belonged to the thieves.

Until the time of Cheryl's death Allan had lived largely on her money, and he now intended to continue to do so once her estate, valued at £450 000, passed to him. He soon sought a new partner to share his life and began dating a former friend and golfing companion of Cheryl's, Jennifer Hughes, herself a wealthy divorcée. Within a month of Cheryl's death they were sleeping together. In November 1998 he gave her a beautiful gold bracelet, although he did not tell her that it had been Cheryl's and had in fact been photographed on her wrist in Luxor—indeed he had reported it stolen. The jewellery Cheryl was wearing when she died had been handed to a holiday company representative who passed them over to Allan, but he claimed that their hotel room had been ransacked and everything of value had been taken. Most of it eventually turned up in a jeweller's shop in Chester, which had paid him £2050 for it. This transaction had taken place only two weeks after his return to the UK.

Cheryl's new will, dated 1993, was exposed as a forgery. The back page, which she had signed and had witnessed by two members of her staff, was genuine, but the earlier pages had been replaced by those written by Allan and attached after her death. This was confirmed when the police impounded his word processor and 67 floppy discs, on which they found a number of different drafts of the three extra pages, all nominating himself as the principal beneficiary. In fact he had typed the new pages in August 1997, more than a year before Cheryl's death. What Allan had not realised was that he had in error spelt her middle name as Leslie instead of Lesley, a slip that told against him when he came to court in 2000. (The original will, in which Allan was not named as a beneficiary, was eventually found on a computer in Cheryl's office.)

Cheryl was to live an extra year because their trip to Egypt originally planned for December 1997 had to be cancelled after a group of Islamic terrorists gunned down 68 tourists at the ancient Temple of Hatshepsut in Luxor on 17 November. Allan's plan to inherit Cheryl's wealth had to be postponed, but he was short of money and looked for another opportunity to obtain it. This arose during a Christmas break in mid-Wales when they visited the Lake Vyrnwy Hotel. During their stay Cheryl had a valuable diamond and topaz ring stolen, and after she returned home in the New Year she offered a £1000 reward for its recovery. She then received a stencilled letter saying that she could have it back on payment of £2150. The stamp on the letter was later examined for traces of DNA and proved to have been sent by Allan. (This fact was not submitted in evidence at his trial.)

When Allan was interviewed, he at first claimed Cheryl had died of a heroin overdose and that she had been an addict for some time. This assertion was disproved when samples of her hair were analysed and revealed no sign of the drug. His next explanation was that he was being framed by agents employed in the arms trade, because he was writing a book exposing their dealings. He stated that he had once been a mercenary and had himself for a time been an arms dealer—but this did not explain why these arms dealers would want to murder Cheryl. Finally, he claimed that he had been framed by Cheryl's first husband, Graham Loseby, who wanted to inherit her wealth.

On 5 February 1999 Allan was arrested and charged with Cheryl's murder. When he appeared before a magistrate on 4 March he said nothing and suddenly appeared ill. An ambulance was sent for and he was kept overnight at a local hospital. The following morning it was clear there was nothing at all wrong with him and he was returned to Walton Prison in Liverpool.

His trial at Liverpool Crown Court began on Monday, 17 January 2000, before judge Mrs Justice (Janet) Smith, whom Allan refused to address in the proper manner throughout his trial and was insolent in his replies to her. The prosecuting counsel was David Steer QC, and the defence was in the hands of Andrew Edis QC. The trial concluded in early March 2000 when the jury found him guilty of murder. Judge Smith sentenced him to life imprisonment, with a strong recommendation that he remained in prison for a very long time. (As soon as Judge Smith had finished the case she rushed off to Manchester to begin the long inquiry into the murders of Shipman, as recounted in Chapter 4.)

8.7 The Tylenol Murders

In 1982 someone in Chicago bought several bottles of a popular painkiller, Extra Strength Tylenol, which consisted of capsules of paracetamol. That person then opened some of the capsules and replaced their contents with about 60 mg of potassium cyanide. He—and it was almost certainly a man—then visited other stores in Arlington Heights, Grove Village, Schaumburg, Winfield and downtown Chicago, and placed the bottles on the shelves, each containing between three and ten poisoned capsules. A series of tragic deaths soon followed. The Tylenol murders were purely random, no motive was ever established, no blackmail ransom was demanded, and no one was brought to justice.

On 29 September, 12-year-old Mary Kellerman of Elk Grove Village, Illinois, complained of a sore throat and a runny nose and her parents gave her an Extra Strength Tylenol capsule. Shortly afterwards they found Mary unconscious on the bathroom floor. She was rushed to hospital where she was later pronounced dead, and the doctors thought she had died of a massive stroke. That same day, paramedics were called to the Arlington Heights home of a 27-year-old postal worker, Adam Janus, and arrived to find him semi-conscious and gasping for breath. He was taken to the local Northwest Community Hospital where attempts were made to resuscitate him, but he died shortly afterwards. It was believed that he had died of a massive heart attack. When they heard of his death his 25-year-old brother, Stanley, went round to Adam's home with his 19-year-old bride, Theresa. They both had head-aches, and each took an Extra Strength Tylenol from a bottle in the kitchen. Within minutes both of them had collapsed on the floor. The emergency services were called and attempted to resuscitate them, but to no avail—Stanley died that evening and Theresa two days later. Doctor Thomas Kim at the Northwest Community Hospital was naturally puzzled by these strange deaths and con-sulted John B Sullivan of the Rocky Mountain Poison Center, who suggested that cyanide might be to blame. Blood samples were sent for testing.

In another part of Chicago four further sudden deaths were re-ported. Mary Kellerman and three members of the Janus house-hold had all died after taking Tylenol capsules, bottles of which had been noticed by the emergency services personnel. The police went to the two homes and retrieved the suspicious bottles. The following day toxicologist Michael Shaffer examined the capsules and discovered that some contained potassium cyanide. Blood samples from the victims confirmed that they had all been poisoned with it.

Extra Strength Tylenol is manufactured by Johnson & Johnson, who immediately undertook a massive recall of the product, totalling 31 million bottles. They issued warnings to doctors, hospitals and wholesale distributors, but it was too late to save the lives of some further victims. Twenty-seven-year-old Mary Reiner of Winfield, Illinois, was recovering after the birth of her son, when she took a Tylenol capsule and died. The same day 35-year-old Paula Prince, a stewardess with United Airlines, was found dead in her Chicago apartment, and the bottle from which she had taken a capsule was later found to contain three further poisoned pills. Another 35-year-old woman, Mary McFarland of Elmhurst, Illinois, was also a victim of the Tylenol murderer.

Soon after news of the random poisonings was broadcast, millions of people either threw away their bottles of Tylenol or handed them in to the police. Hospitals around the country admitted patients who thought they might have been poisoned, but there were no further cases. On October 2 1982, another contaminated Tylenol bottle was recovered by police from a drug store in the Chicago suburbs. Johnson & Johnson inspected the factories where the Tylenol was produced to see whether the cyanide could have found its way into the capsules during the production process, but found nothing. They offered a $100 000 reward for information leading to the arrest and conviction of the murderer. The FBI and the FDA now knew that someone had deliberately put potassium cyanide into the capsules. The most obvious suspect was someone with a grudge against Johnson & Johnson—but who? The killer had made no attempt to contact the company.

They knew it was little use trying to trace the source of the potassium cyanide since it was so widely used in gold and silver mining, in steel plating, for film processing, and in chemical manufacture. However, there were one or two suspects. One was a 48-year-old employee in a warehouse which supplied Tylenol to two of the stores from which the poisoned bottles were bought— and he was known to have an interest in chemistry, although only at an amateur level. When questioned he admitted to having at one time worked with cyanide, and in his apartment investigators found guns, an airline ticket to Thailand and a book about poisons, but in themselves these proved nothing. He could not be linked to the murders and in the end was charged only with illegal possession of firearms and sent to jail.

Another suspect was James W Lewis, who had attempted to extort $1 million from Johnson & Johnson. He was eventually arrested in New York, and ended up in jail for a series of frauds, but not for the Tylenol murders. As for Tylenol itself, it was now relaunched in tamper-proof packaging and within weeks it was re-established as a leading brand—but this did not prevent it being targeted again. While visiting her boyfriend in New York in February 1986, 23-year-old Diane Elsroth took two Extra Strength Tylenol capsules from a seemingly tamper-proof bottle and died almost at once of cyanide poisoning. Her boyfriend was lucky, because the bottle contained three further poisoned capsules. Tylenol was again recalled from stores in the area and another bottle containing poisoned capsules was found at Woolworth's in Westchester County, New York. The killer had somehow been able to make the packaging look as though it had not been tampered with. He too was never found, and there were no further incidents.

Approximately four months later a further killing occurred, this time involving another popular painkiller, Excedrin. On the morning of 11 June 1986, Sue Snow, a 40-year-old Washington State bank manager, took two Extra Strength Excedrin capsules before her shower. Forty minutes later she was found unconscious on the floor and was rushed to a nearby hospital, but died a few hours later. An autopsy and blood analysis revealed that she had died of cyanide poisoning. Laboratory tests confirmed the presence of cyanide in some of the capsules at her home and the manufacturer, Bristol-Myers, put out a national recall of the product on 16 June.

The following day, Stella Nickell rang the police to say that she suspected that her husband Bruce might also have been poisoned. He had suddenly dropped dead on 6 June after taking Extra Strength Excedrin, but his death had been attributed to complications of emphysema. Tests on his blood now confirmed the presence of cyanide, and adulterated capsules were discovered in an Excedrin bottle at his home. Two more bottles of Excedrin Extra Strength Capsules from stores in Auburn and Kent, Washington State, were also found to contain cyanide. In this instance, however, Stella Nickell herself turned out to be the murderer, and this was proved when the potassium cyanide in the capsules was found to contain minute specks of green crystals. These were shown to be an algaecide known as Algae Destroyer, used for aquaria and fishponds, and it was deduced that the murderer had probably mixed the cyanide in a container that had previously held this product. Stella Nickell not only owned a fish tank but had also bought Algae Destroyer, and she had ground this in the pestle and mortar which she subsequently used to grind the cyanide.

Nor was her motive difficult to deduce. She had taken out three life insurance policies on her husband the previous year worth a total of $71 000, plus an additional $100 000 if his death was accidental. Stella thought that since he had been the victim of a random murderer she would become eligible for the extra payment. However, a lie detector test which she took on 18 November 1986 convinced the police that she was guilty. They also had a statement from Stella's 27-year-old daughter, Cindy Hamilton, who came forward to say that her mother had often talked about killing Bruce, and had already unsuccessfully tried to poison him with digitalis, from foxgloves. She said her mother had learnt about cyanide at the local library, and their records showed that she had in fact borrowed a book entitled *Deadly Harvest*. On the pages devoted to cyanide her fingerprints were found. Stella was charged with the deaths of Sue Snow and Bruce Nickell on 9 December 1987 and her trial began in April 1988. A month later, on 9 May, she was found guilty and sentenced to 90 years in prison.

Another copycat murder bearing remarkable similarities to the Tylenol and Excedrin poisonings occurred in 1991, again in Washington State, at a town named Tumwater. Joseph Meling (31) had insured his wife for $700 000 and then bought 1 lb (454 g) of sodium cyanide for $11 from the Emerald City Chemical Company in the town of Kent, about 80 miles from Tumwater. He used a false name, supported by a fake driver's licence as ID. He also bought packs of Sudafed decongestant capsules[ix], which he knew his wife would take if she had a blocked nose, and he replaced the contents of some of these with cyanide. One pack of Sudafed he kept at home, the others he put back on the shelves of stores along Interstate 5 Highway going northwards. He wanted to make it appear that a serial poisoner was on the rampage. His wife Jennifer took one capsule on 2 February 1991, became ill, and Meling called for medical help. She was rushed to hospital, and the doctors

---

[ix] The active ingredient in these was pseudoephedrine hydrochloride. This has since been replaced, because it can easily be converted to the street drug, methamphetamine ("crystal meth").

managed to save her life. They were however puzzled by what might have caused her condition and Meling himself even suggested it could be cyanide.

Others were not so lucky. Kathleen Daneker (40) of Tacoma died on 11 February and Stanley McWhorter (44) of Lacey on 18 February. Tacoma is 20 miles from Tumwater and Lacey about 10 miles. Their deaths seemed suspicious, and blood tests revealed they had been poisoned with cyanide. It was suspected that another random murderer was at large and had targeted Sudafed. Nine hundred packs of this antihistamine, totalling a quarter of a million capsules, were collected and analysed. It was clear from certain packs that something was amiss because the tamper-proof seals easily came apart. Meling was arrested and brought to trial, and found guilty of six cases of product tampering, two counts of perjury, two murders, and one of attempted murder—that of his wife. He was sentenced to life imprisonment without parole. Although it was of little consolation, or help to his case, his wife told the court that she had forgiven him.

8.8 Analysis of Suspect Capsules

When the contents of tens of thousands of suspect capsules need to be analysed, how do chemists approach the task? Today they use more advanced techniques, but at the time of the Tylenol scare in 1982 they relied on visual inspection of the capsules, looking for any bearing the stress marks which showed they had been reassembled by hand. The contents of the suspect capsules were then examined under a microscope, and it was possible to distinguish the crystals of sodium cyanide from those of paracetamol. These could then be analysed fully to determine how much cyanide was present. The analysis could be carried out in a number of ways, for example by determining the amount of sodium present using atomic spectroscopy. Unfortunately at the time of the first Tylenol murders some of the poisoned capsules had been contaminated by the medical examiner who first analysed them.

Today in the USA, the FDA has its own forensic chemistry laboratories in Cincinnati where 30 chemists are employed to examine suspect foods and other products that may have been accidentally or deliberately tampered with. The chemists have a battery of techniques to assist them. Contaminants can be isolated using thin-layer, gas, liquid, or ion **chromatography**, or by **capillary electrophoresis**. They can then identify the separate components using instrumental techniques such as **mass**, **atomic** and **infrared spectroscopy**. The FDA laboratories also have three biologists and they can employ an **electron microscope** in addition to the more normal microscopes, and with it can identify different particles of the minutest kind.

The FDA is now able to trace the source of a sample of sodium cyanide by identifying traces of other elements present, which can be of the order of parts per trillion. In effect, each batch of sodium cyanide has its own 'fingerprint' indicating its origin. Not surprisingly, copycat murders of the Tylenol type are now a thing of the past—but keep your fingers crossed.

CHAPTER 9

Paraquat and the Poisoned Gravy

A word shown in **bold** *indicates that further information will be found in the Glossary*

9.1 Weeding Out the Unwanted

Paraquat is an excellent weedkiller, widely used in farming around the world to boost crop yields. It is also an insidious poison and for a time was used in domestic murders. Today it is virtually impossible to use it as a murder weapon since other chemicals are now added to warn of its dangers. However, when paraquat was first introduced in the 1960s these additives were not considered necessary until it became apparent that paraquat was causing accidental deaths, as well as being used to commit suicide. A few people were even using it to weed out unwanted individuals. Paraquat poisoning was difficult to diagnose and almost impossible to treat.

For a few short years paraquat appeared to be the perfect poison, and there were some classic murder cases involving it. Steven Catlin used it to dispose of various of his wives, as well as his parents. Susan Barber resorted to it when she decided to murder her husband. He had gone on a fishing trip one Saturday morning, but bad weather led to it being cancelled and he returned home earlier than expected to find his wife and her lover in bed. He reacted predictably and resorted to violence, directed towards both of them. Her revenge was to poison him, and she almost got away with it, thanks in part to a mistake made at the hospital where her husband died. 'Nympho' Susan, as she subsequently referred to herself, consoled

Molecules of Murder: Criminal Molecules and Classic Cases
By John Emsley
© John Emsley, 2008

herself with new lovers, wild parties and porno movies. And so it continued, until one day there was a knock at the door . . .

9.2   Paraquat as a Defoliant

Paraquat is the common name for a chemical containing the 1,1'-dimethyl-4,4'-bipyridinium ion. It is sold commercially as a concentrated solution of the dichloride, which is very soluble in water, at up to 700 g per litre.[i] Crystals of paraquat are colourless and odourless, and are hygroscopic, in other words they absorb water from the atmosphere and become moist.

The molecule was first synthesized in 1882 and given the name methyl viologen. It was used as an indicator in redox reactions and signalled the end point of a titration by turning bright blue in the reduced form. (A simple chemical trick is to turn the blue solution colourless merely by shaking it in a half full bottle. What is observed is methyl viologen becoming oxidised by the air in the bottle.) That methyl viologen had potential as a powerful herbicide was not appreciated until 1955, when chemists at Imperial Chemical Industries (ICI), in the northwest of England, tested it. The company began manufacture in 1961 and it was launched as a broad spectrum herbicide the following year under the name paraquat. Farmers and horticulturalists used it to kill the weeds which reduced crop yields by competing for water, nutrients and light. It was also ideal for keeping orchards and woodlands free from undergrowth without affecting the trees themselves. Householders and gardeners used it to clear paths and patios, and municipal authorities sprayed it in parks and other public places to keep the weeds down.

The ICI chemists who discovered the herbicidal properties of paraquat were naturally intrigued by the potency of this simple compound and developed a range of analogues with slightly modified structure. Examples included **diquat** and **diethamquat**, and researchers in other companies produced **cyperquat** (Gulf Chemicals) and **difenzoquat** (BASF).[ii] Paraquat, however, remained in pole position.

Although patent protection for paraquat expired many years ago, ICI (now Syngenta) remains the world's main producer. The company has production sites throughout the world, including a new one

---

[i] Paraquat is slightly soluble in alcohol and practically insoluble in other solvents.
[ii] The 'quat' element signifies that these are all **quaternary ammonium** salts.

in China, north of Shanghai, which produces 6000 tonnes of paraquat annually. The product is approved for use in over a hundred countries, the main markets being the USA, Japan, Malaysia, Thailand, Mexico, France and Brazil. It remained the best-selling herbicide worldwide until glyphosate (Roundup) arrived on the scene.

Paraquat is sold under a variety of trade names, such as Gramoxone, Firestorm and Weedol, and in the USA it is also known as Cyclone. It is marketed as a concentrated solution, ranging in strength from 24 to 36%, and diluted as necessary before use. Cyclone is used in cotton plantations as a defoliant prior to harvesting and is applied at around 1 litre per hectare. It is also used for weeding land before planting crops such as alfalfa, asparagus, leeks, onions and sugar beet, in which case the dosage is 0.5 litre per hectare. Its more general use around the world is for controlling undergrowth in plantations of bananas, coconuts, coffee, olives, palm kernel oil, rubber and tea. It can also assist in drying crop residues after harvesting, particularly in plantations of pineapples, soya, sugar cane and sunflowers. This helps in clearing the ground before either burning the unwanted weeds or, better still, harvesting them for use as fuel. In pineapple groves paraquat enables the land to be cleared after about a month, rather than the three months otherwise required. Paraquat is increasingly used for no-till agriculture, avoiding the need for ploughing, which is now recognised as a major factor in soil erosion and in releasing considerable amounts of carbon dioxide from the soil into the atmosphere.

In some countries paraquat has even been used deliberately to kill crops, for instance in Mexico, where the object was to destroy the fields of marijuana which supplied users in North America. This was carried out with the help of US agencies. Furthermore, in an attempt to deter people from smoking marijuana, those who organised the programme broadcast a warning that the spraying rendered the leaves unsafe to use. However, chemists at the US Environmental Protection Agency defused the anti-drug propaganda by denying that there was any such danger, noting that when marijuana was smoked any paraquat residues would rapidly be pyrolysed to harmless components. The main danger was in fact to those involved in spraying the illegal crops.

Paraquat is rapidly absorbed by the leaves of plants within minutes of application, which has the advantage that even if it rains soon afterwards it is not washed off. Effective contact with leaf tissue is improved by the inclusion of a surfactant in the formulation. Anionic surfactants such as alkylbenzene sulfonates are not compatible with paraquat since this contains a positively charged cation, and a nonionic surfactant such as **Lissapol N** is used instead.

Paraquat does not kill plants by poisoning them directly. It is the changes it causes in the leaves into which has been absorbed when these encounter light and oxygen. Under paraquat's influence these form **superoxide** free radicals which are highly dangerous and will disrupt any other molecule they come in contact with. (Paraquat kills humans by poisoning them directly and it does this by attacking key enzymes from which it abstracts electrons and thereby making them malfunction.) Paraquat's attack on plants occurs at the leaf's chloroplast, a special cellular compartment found in the cells of all green plants and where photosynthesis takes place. This is the process that converts carbon dioxide and water to the glucose which is needed to make components such as cellulose, the fibrous material which supports the plant. Photosynthesis requires the capture of a photon of light to energise an electron which is then used to drive a complex series of reactions. Paraquat interferes with the chloroplast's first stage in which the active electrons are generated. These are lured away by the paraquat to form a free radical ion. This now passes the electron on to an oxygen molecule, O_2, to form a superoxide free radical, $O_2^{\bullet-}$. Relieved of its newly acquired electron the paraquat is regenerated and ready to repeat its electron-snatching theft again. Meanwhile the deadly superoxide free radical looks around for an easy victim and finds one in the double bond of the unsaturated fatty acids that constitute part of the cell membrane. Once these are chemically changed then the membrane loses some of its structural integrity and water begins to leak from the cell. This loss of essential fluid eventually results in the leaf dying. Paraquat-treated plants wilt within hours under direct sunlight although this water loss may take several days when the sky is overcast. The speed of cell destruction is too rapid to be remedied by more sap, cell repair, or new growth.

The reactivity of paraquat is confirmed by its unexpected ability to attack aluminium. This metal is normally protected from corrosion by a layer of oxide on the surface, otherwise it rapidly corrodes. Paraquat strips away the oxide layer, leaving the metal open to attack.

9.3 Paraquat as a Poison

Paraquat first achieved notoriety as the result of accidental poisoning, typically when children drank the weedkiller stored in old soft drink bottles. Swallowing a mouthful spelled almost certain death unless medical help was immediately available. Once its poisonous nature became recognised paraquat attracted the attention of those intending to commit suicide. It also began to be used as a murder weapon. A dose as small as a teaspoonful, slipped surreptitiously into a drink such as Coca Cola, was all that was required and after a few hours had elapsed no amount of medical help would save the victim.

Safety measures were introduced in the 1970s to reduce accidental poisoning, and also to discourage acts of suicide or murder. A vile smelling chemical, graphically described as a stenching agent, was included in the formulation in 1975 and an emetic was added in 1977. A further preventive measure was to colour it with a green or blue dye, to avoid its confusion with dark-coloured drinks such as colas, black coffee or similar beverages.

In Japan the bitter substance, **Bitrex**, was added, to reduce the growing number of older men who were poisoned with paraquat. Retirement in Japan often meant that a man spent much of the day at home. Such men were referred to as 'wet leaves' by their spouses since they were equally difficult to get rid of. Paraquat was a means of solving the problem, often on the part of a wife to dispose of her spouse, but sometimes by the depressed husband to kill himself—some men are known to have drunk as much as half a litre of the weedkiller. Bitrex virtually put an end to these murders and suicides, as the statistics confirm. In Japan a total of 346 cases of paraquat poisoning took place during the years 1998–2002, of which 242 were attempted suicides, 97 were occupational accidents, and only 7 were murders. Of the 346 cases 90 people died.

In the dilute form used for crop spraying paraquat is not acutely toxic. The greatest risk of accidental poisoning is during the dilution process. In 1992 an agricultural worker in the UK died after being splashed in the face with paraquat when he dropped an open container. In Guápiles, one of the main plantation regions of Costa Rica, there were 284 such accidents involving paraquat over the years 1988–1990, but happily none was fatal. Those involved in spraying the diluted material can however be affected, especially if the herbicide comes in direct contact with the skin. Continued exposure to it by plantation workers can also affect the eyes, nose and finger nails. The affected skin may show symptoms ranging from a mild rash to blistering and ulceration, and in severe cases the outer layer of the skin peels away. Severe exposure to the hands causes nail loss, and splashes in the eye can result in inflammation of the eyelids and impaired vision.

In its pure state paraquat has an acute oral **LD$_{50}$** in rats of 157 mg per kg, which classifies it as moderately hazardous, but much less toxic than for example arsenic. The lethal dose of paraquat by oral ingestion is around 35 mg per kg of body weight, about 2.5 g for an average 70 kg adult, ten times the fatal dose of **arsenic**. The World Health Organization has decreed that an acceptable daily intake of paraquat is 0.005 mg per kg body weight, a mere 0.35 mg for a person of average weight.

The toxicity of paraquat depends very much on the route by which it arrives in the body, and also how quickly it is treated. If it is swallowed the best treatment is to flush the stomach with an absorbent material such as fuller's earth or activated charcoal. Timely intervention can be effective, but if this is delayed by several hours death is virtually inevitable, even though it may be slow in arriving. There have been remarkable cases of survival—or almost survival. Some of those attempting suicide have remained alive for up to five weeks. According to the US Centers for Disease Control, ingesting paraquat can damage the lungs, liver, oesophagus, heart and kidneys. The lungs selectively accumulate paraquat and it is there that it does most damage, as we might expect knowing its relationship with oxygen. This results in pulmonary oedema and other lung damage, leading to fibrosis and reduced flexibility within the lung, which hinders uptake of oxygen by the blood. It is this which leads to death. Paraquat is thought to be mistaken for a peptide by the body as a consequence of its molecular structure, particularly the positioning of its two nitrogen atoms—see Glossary. Some commercial paraquat analogues contain the nitrogens in different respective locations and as a result are much less toxic, as is the case with **difenzoquat**, in which the nitrogen atoms are adjacent to one another.

The conditions under which paraquat is used in many developing countries still leave agricultural workers at risk. Safety recommendations on the label are either not understood or are simply ignored. More important are inadequate protective clothing, lack of training and the absence of washing facilities. A study of small farmers in Kenya found that protective clothing was almost never worn, which is hardly surprising as the cost of a pair of rubber gloves was equivalent to a day's wage. There have been some puzzling incidents, such as that in the Dominican Republic, where paraquat is widely used on banana, coffee and cocoa plantations and to keep down weeds on verges of roads. During October 2002 a small amount of herbicide spray drifted into a textile factory and 150 of the workers complained of headaches, nausea, dizziness, dehydration and lack of energy. That so little paraquat could have had such an effect does however suggest that some degree of mass hysteria might have been involved.

There are also concerns that the widespread use of paraquat could be damaging to the environment. According to the EU Commission, however, it poses hardly any threat, as it is bound tightly to clay particles in the soil and is not washed off. Research has shown that paraquat residues in the soil vary between 22 and 58 mg per kg (22–58 ppm), and long-term field studies reveal that these degrade at the rate of 10% per annum. In sandy soils with low organic content paraquat may be more readily released into ground water, where it can pose a threat because aquatic plants are able to concentrate paraquat to high levels. Studies have suggested significant tadpole mortality in contact with paraquat-contaminated aquatic plants. There is also evidence of toxicity to fish, varying with the size and species involved.

In 1983 the German BBA[iii] reported that repeated treatment with paraquat could lead to its accumulation in the soil and cause damage to crops, and this body declined to re-register paraquat. The evidence was subsequently challenged in the courts and in 1992 it was ruled that, while the BBA action had been justified, a re-formulated product with a lower level of the herbicide (10%) could be approved for use, although only in designated areas. Application to crops was limited to once in four years, and only in areas free from erosion or flooding.

Paraquat is known to be lethal to domestic animals. For dogs, the fatal dose is 25–50 mg per kg, and for cows and sheep it is

---

[iii] Biologische Bundesanstalt für Land und Forstwirtschaft (Federal Biological Institute for Farming and Forestry).

50–75 mg per kg. Deliberate poisoning of animals has been reported, and in 1990 in Australia a flock of 700 sheep died after
paraquat was added to their drinking troughs. Wildlife can also be
affected, some species being particularly vulnerable. In the UK in
1976 and 1990 paraquat was employed to kill grassy stubble and
resulted in the deaths of hundreds of hares. Paraquat is however
less toxic to birds than to mammals.

In most developed countries paraquat is strictly controlled with
regard to how much can be applied, where it can be used, and the
method of application. Paraquat is banned completely in countries
such as Finland, Norway, Sweden and Kuwait, although none of
these are countries which rely on agricultural exports. Its use is
restricted by law in Belgium, Chile, Germany, Indonesia, Korea and
in the Caribbean, and in the USA only certified personnel are
allowed to purchase it. As the result of its use in poisonings, of
which most were suicides, Austria and Hungary banned it altogether. There are no restrictions on the use of paraquat in the
UK, except that a licence is required to purchase it, and there have
been few incidents of accidental or deliberate misuse.

The most obvious effect of paraquat in humans is lung damage,
even when the poison has been swallowed rather than inhaled, and
this condition has become known as 'paraquat lung'. The symptoms are shortness of breath, a sore throat and vomiting. The most
likely cause of death is pulmonary fibrosis leading to respiratory
failure. The lungs are particularly sensitive and it is essential to see
that the patient has a clear airway, but oxygen should be administered only when there is difficulty in breathing. It has been
suggested that in this situation the addition of nitric oxide to the
oxygen may be beneficial.

Paraquat poisonings in South Korea, where a 25% solution of
paraquat dichloride was available as a weedkiller, were analysed
for the period 1992–1996 and 106 cases were noted. The level of
the poison in the plasma of those who died within a day of taking
the poison varied widely, between 2 and 636 μg per ml (ppm).
Those who survived beyond a day, but died within four days, all
had between 1 and 25 ppm in their plasma. The prognosis was that
death within a day of drinking the weedkiller was almost certain if
the level exceeded 25 ppm. In 1983 a person attempting to commit
suicide was stated to have drunk more than 20 g of paraquat and
died within 18 hours, the main symptom being metabolic acidosis

and hyperglycaemia rather than damage to the lungs. The post-mortem showed some liver damage, but the other organs of the body did not appear to have been affected.

In 2005 a group of doctors at the University of Medicine in Cheonan, South Korea, reported that ethane (C_2H_6) was exhaled by those suffering from paraquat poisoning, formed as the result of attack on the lungs by the peroxide radical. However, they found that the amount of ethane emitted did not correlate with the quantity of paraquat in the body, and it therefore could not be used to predict the likelihood of survival. There is no antidote to paraquat, although some doctors have administered large doses of antioxidants, in particular vitamin C and vitamin E at 4000 mg and 250 mg per day, respectively, but there is no proof that these have much effect.

9.4 Murder by Paraquat

In the early years, when paraquat was readily available and con-tained no protective chemicals, it came to the attention of some individuals that it might well be used to take the lives of those they no longer wanted around. Even after protective chemicals were added to paraquat there was still some of the older formulation available, and these could be given to the victim who would be unaware of being poisoned. Two such cases, one in the US and one in the UK, are classics of their kind.

9.5 The California Dreamer

Steven David Catlin was adopted as an infant in 1944 by Glenn and Martha Catlin, who lived in Kern County, California. Forty years later he murdered his mother with paraquat, a crime for which he was found guilty and for which he awaits execution, al-though this is now regarded as unlikely to take place.[iv] He may well have killed his father and two of his six wives in the same way.

Catlin's life of crime began during his teenage years when the family moved to Bakersfield, a city of a quarter of a million in-habitants 40 miles north of Los Angeles. Catlin dropped out of high school and with no regular job he resorted to forging cheques to

---

[iv] California reinstated the death penalty in 1977 and there are currently more than 600 prisoners awaiting execution, but these are unlikely to meet this fate because public sen-timent is strongly against it. In fact only nine prisoners have been executed in the past 30 years.

fund his addiction to drugs. He was arrested, and in 1963 spent 40 weeks at a California Youth Authority camp. Not that this corrective training seems to have made him see the error of his ways.

Catlin's sex life was also in need of discipline. His first marriage was beset with domestic problems due to his abuse, both of drugs and his wife, and she finally left him. Catlin soon found a new partner, whom he married bigamously in 1966 under a false name. That marriage lasted for almost four years, the reason being that Catlin was serving a three-year prison sentence for stealing a credit card to finance his drug addiction. In his cell, Catlin could only dream of the summers of love for which California became famous in the late 1960s.

When he was released from jail Catlin did the honourable thing and divorced his first wife, before legally marrying his second wife under his real name. But it provided little benefit for the relationship and after ten months the couple separated. Catlin now sought a third wife and married Edith Ballew, but this marriage lasted only eight months. Nevertheless Edith continued to take an interest in her former husband's affairs and it was due to her persistence that he was ultimately brought to justice.

His fourth marriage, to Joyce Adeline, survived rather longer but ended tragically with her death when she was only 40 years old. She worked in the welfare department of the local authority, and he was a garage hand or worked in service stations, but he was always in need of cash. In the middle of April 1976, Joyce was admitted to Bakersfield's Mercy Hospital with a severe case of what appeared to be a flu-like infection, and breathing was difficult. For a time she seemed to improve but then took a sudden turn for the worse, and died on May 6 of what was diagnosed as pneumonia. Her death certificate gives the cause of death as acute respiratory failure due to an unknown microorganism. Husband Steve chose to have her body cremated, thereby destroying the evidence that he had almost certainly poisoned her with paraquat.

What he did not know was that the hospital had actually suspected foul play, and it was even suspected that she might have died of paraquat lung. They had therefore carried out an autopsy and had retained some tissue samples. At that time there was no analytical test able to confirm the presence of paraquat if the poison had been given more than three days previously, still less to quantify the amount present in her tissues. The samples taken from her body were kept frozen, however, and were finally sent for forensic analysis eight years later, in November 1984, when such analysis was feasible and evidence was needed to support the charge that he had murdered her.

A year after Joyce's death, in May 1977, Catlin married his fifth wife, Glenna Kaye, and they moved to Fresno where he gained employment at a large local garage. There was no doubting his experience with cars—he had worked at one time as a member of the pit crew for a racing car enthusiast—and he was quickly promoted to the role of manager, with around 40 people working under him. Despite his rise in the world, Catlin was still short of money. His answer may have been to murder his father Glenn with paraquat, because his death also was consistent with its symptoms. Glenn died suddenly on October 28 1980 and his lungs were found to be congested with fluid, although at the time this was regarded as the result of his lung cancer, which had been diagnosed earlier. He too was cremated. Whatever money his father's death brought Catlin was soon spent, however, and Catlin now resorted to stealing from his employers, who eventually noticed that a number of expensive automotive parts were missing. At first they did not know whom to suspect, but they carried out a background check on some of their employees and this revealed Catlin's police record. They confronted him and asked him to resign, which he agreed to, on condition that no charges were brought against him.

Catlin was now in serious financial difficulties, and decided that it would help matters if his fifth wife were to die. He slipped her a dose of paraquat on the day she left for a holiday in Las Vegas with her mother. Soon after their arrival on Friday, 17 February 1984, Glenna began to feel ill and they returned to Fresno, where she was hospitalised and diagnosed as having fluid in the lungs. The doctors were puzzled by her condition, and were still trying to discover the cause when she died on 14 March. Glenna Kaye had carried life insurance, and Catlin now collected $57 000.

He quickly acquired yet another fiancée, whom he had actually met while visiting his dying wife in hospital, and they were soon married. To add to Catlin's series of family tragedies his poor mother, Martha Rose, passed away on 8 December at her Fresno home, aged 79. Her death was attributed to a stroke, although we now know that it occurred shortly after a visit from her son. Catlin wanted her body to be cremated, but this was postponed until an autopsy had been performed. One person who continued to take an interest in the deaths which appeared to dog Catlin's steps was his third wife, Edith, and she now voiced her suspicions to the local sheriff. Tissue samples from his mother's body, together with those from the bodies of his fourth and fifth wives, were analysed and showed that they had been poisoned by paraquat. Catlin was questioned and arrested.

He was put on trial at Monterey in May of 1986 for the murder of Glenna Kaye. Evidence was given that Catlin, who had worked on a farm in the 1970s, had had access to paraquat and had in fact mentioned that it could be used as a poison. Police obtained a bottle of paraquat bearing a 1977 date from the garage which Catlin had shared with Glenna Kaye's father. At that time the weedkiller was sold in America without the additives giving it a distinctive colour or smell. The jury found Catlin guilty, and he was sentenced to life imprisonment without parole.

Back in Bakersfield another prosecution was being planned, and this time it was one which carried the death sentence. In 1990 he was charged with poisoning his mother in 1984 and his wife Joyce in 1976, each with paraquat, although the jurors were not told that Catlin had previously been convicted of poisoning at the Monterey trial. Evidence for the prosecution included details of the financial benefit he derived from the deaths. In particular, he was the sole beneficiary of his mother's estate and the court heard testimony that he had been fearful that his mother might disinherit him and leave her estate to charity, since she disapproved of his multiple marriages and divorces. The jury found him guilty of both murders and Judge Lewis King sentenced him to death. As tends to happen with serial poisonings, the case attracted widespread attention and was even made into a television movie in 1993, called *Poisoned by Love: The Kern County Murders*. Steven Catlin was portrayed by Harry Hamlin, better known as star of the TV series, *LA Law*.

Catlin appealed against his sentence on the grounds that his conviction for the death of Joyce related to a crime committed before the Californian law had reinstated the death penalty, but this argument failed to convince Chief Justice Ronald M. George. Catlin's defence lawyer additionally reasoned that it was unfair not to have brought the prosecution against him earlier in respect of Joyce's death. That too was waved aside. Against the appeal the prosecution presented testimony from a jailhouse informant that

Catlin had made an attempt to contract someone to threaten his third wife in order to prevent her making accusations that he was a poisoner. Also while in jail, he foolishly admitted to another prisoner that he had poisoned his wives.

Judge George summed up: "Because of limitations in forensic science and because of the manner in which Joyce's tissue had been preserved, it would have been extremely difficult or impossible to make out a case against the defendant at or near the time of the murder." The second trial dealt with two murders, those of his mother and his wife Joyce, and was in the judge's view appropriate due to the similarity between the two murders, and evidence regarding the related death of Glenna Kaye was admissible for the same reason. The judge added: "Paraquat poisoning is rare, and its occurrence with respect to two close relatives of one person is unlikely to be a matter of chance or to be the result of a spontaneous [suicidal] impulse. When evidence of a third instance of the same type of poisoning is introduced, as it properly was in the present case, the inference regarding a common design or plan becomes very strong."

9.6 "I've Made You Gravy for Your Pie"

In the UK there is a popular but misguided belief that women in Essex are blonde oversexed airheads, and numerous jokes have been told about them. They are supposed to be fond of the good life and to make ideal partners for football stars. They are the archetypal WAGs (wives and girlfriends) so beloved of the tabloid press. Was the subject of the next poisoning case just such a woman? Possibly.

The subject of this story did not achieve WAG status, but she did latch on to Michael Barber. He was born in 1946, and by the time he was 24 he had a criminal record for stealing cars. He was also guilty of making Susan, his 17-year-old girlfriend, pregnant—at least that's what she told him, but it never dawned on him that the baby girl might not actually be his. He and Susan were married in 1970 and lived at 29 Osborne Road, Westcliffe-on-Sea, Essex. They went on to have two more children. Michael was again in trouble with the police, when he was charged in 1972 with indecently assaulting his six-year-old niece. He was not an easy person to live with and on two occasions Susan left him, but each time she returned. She too led a life of less than domestic sobriety. Unbeknown to Michael, she was engaged in an affair with a teenage neighbour, Richard Collins, whose family lived in the same street only three doors away. That liaison began in 1980.

Susan and Richard enjoyed their sexual romps early in the mornings, just after husband Michael had left for work. Michael was employed as a packer at the Rothmans cigarette factory six miles away in Basildon. Lover Richard would call at number 29 on his way to work, and if there was an opportunity to join Susan in bed he did so. They also arranged to meet at other venues and other times, and all the while Michael Barber was unaware of what was going on—until one fateful Saturday in March 1981. Michael went fishing with a friend and had left home at about 4 am. The coast was clear, and early that morning Richard went round to number 29.

What the lovers did not realise was that the weather had become so windy that Michael and his friend decided to cancel their trip. The classic comedy situation which has featured in so many plays and films then unfolded. The front door of number 29 suddenly opened, and the cry, "I'm back!" rang up the stairs. Richard leapt frantically out of bed and began to get dressed, only to be confronted by Michael who, realising at once what had been going on, began to punch him. Richard fled back home, quickly packed a bag and went to live with his brother in the nearby seaside resort of Southend.

Michael also gave his naked wife a number of blows, one of which struck Susan on her ear with such force that it needed medical attention. The doctor who attended her injuries advised them to seek marriage guidance counselling, and Susan said she wanted to save the marriage and agreed to behave herself in future. But although she and Richard did not meet for the time being, she continued to send him notes through a go-between. Susan had thought of an alternative way to renew the relationship—she would poison Michael. She knew the ideal agent: there was paraquat weedkiller in the shed at the bottom of the garden.

On Thursday, 4 June, Michael went to work but had such a painful headache that he attended the factory clinic and the paramedic gave him some tablets. He had in fact been unwell for several days. The following day there was no improvement, and by Saturday he had developed severe stomach pains and was violently sick. He also had an intensely painful sore throat. The doctor who was called diagnosed food poisoning and prescribed an antibiotic, but Michael became steadily worse, so much so that when the doctor called the following day he rang for an ambulance and had Michael admitted to Southend General Hospital. Nothing the doctors and nurses did there could prevent his condition deteriorating, and he was placed on a ventilator in intensive care. Tests were carried out and he was diagnosed as suffering from Goodpastures syndrome, a rare condition affecting the kidneys, which in Michael's case were clearly starting to fail. More expert help was needed, and on 17 June he was transferred to Hammersmith Hospital in London where specialist renal treatment was available, but he remained critically ill.

The doctors were unable to identify Michael's illness, but they eliminated the possibility of both Goodpastures syndrome and Legionnaire's disease, which they had also suspected might be the cause. Then a consultant specialising in respiratory medicine suggested that he might be suffering from paraquat lung. Could he have been exposed to this material? Michael said it was not possible. Nevertheless the consultant instructed that blood and urine samples should urgently be sent for analysis at the National Poisons Reference Centre in south London. Due to an oversight by the nursing staff this was not carried out, and when the consultant later asked for the results and the mistake was realised, they covered it up by saying that samples had been sent but the results had come back negative. Even if they had known that Michael had been poisoned with paraquat, it was too late to save him. Michael Barber died on 27 June 1981.

Michael's death certificate gave the cause of death as cardiac arrest, renal failure and bilateral pneumonia. A post-mortem examination was carried out by Dr Peter O'Brien and Professor David Evans, who were puzzled by what they found. They removed Michael's major organs and preserved them in **formalin**. The autopsy report came to no firm conclusion, although the two pathologists expressed the opinion that their findings were consistent with paraquat poisoning.

Michael Barber was cremated at Southend Crematorium on Friday, 3 July 1981. Richard Collins attended the ceremony, and while both he and Michael's widow put on a public display of grief, they were weeping crocodile tears. That weekend Collins moved in with Susan Barber but their rekindled relationship lasted less than two months. Susan had found herself a new lover, and Richard was out.

In October 1981 Susan Barber inherited her husband's estate, the
major financial benefits of which were a lump sum of £15 000 and
an annual pension of £1700.[v] When Susan told her latest fancy
man that Collins had made off with some of this money he confronted him, demanding it back—but of course it was already
spent. The new lover then brutally assaulted Collins, for which he
was arrested and eventually jailed.

Susan invested some of her new wealth in a CB radio and gave
herself the code name Nympho, entertaining her fellow CB
enthusiasts with raunchy chitchat. She found yet another new
lover, to whom she was attracted by his mysterious past and supposed familiarity with black magic. In actual fact he was already
known to the police as a drug dealer. The modest terrace home at
29 Osborne Road became the scene of late night drinking parties
and porno movie shows. Her sexual demands were insatiable, or
this was what one of her lovers later told a tabloid newspaper,
claiming that they had had sex several times a night.

Meanwhile Michael's strange death continued to puzzle the
doctors at Hammersmith Hospital. In September 1981 they examined the histology slides of the samples of his organs and concluded
that he had indeed been poisoned, probably with paraquat. In
January 1982 the doctors and consultants held a meeting and decided that matters should be taken further. By this time they were
aware that chemists at the National Poisons Unit had not in fact
analysed Michael's blood—because they had never received it.
Samples of the preserved tissue were sent to ICI, the manufacturers
of paraquat, and their analysts came back with positive evidence
that paraquat was present in every sample. On 15 February 1982,
the Hammersmith coroner was informed of these findings and he
instructed the police to investigate the suspicious death of Michael
Barber.

Detective Chief Inspector John Clarion took charge of the case.
His officers were sent to interview all of those involved in treating
Michael, and further forensic analysis was carried out on his
remains. These investigations had to be kept secret from some of
the local police in the Westcliffe-on-Sea area, because Collins' sister
was a policewoman and his mother had also worked in the police

---

[v] Equivalent values in 2008 would be around £40 000 for the lump sum and £4500 for the
pension.

canteen and knew many local police officers. The interviews were carried out during February and March 1982, neither Susan nor Collins being aware that their days of freedom were numbered. When all the medical and forensic evidence had been assembled a leading pathologist, Professor James Cameron of the London Hospital, was asked to examine it and give his opinion. He said that the evidence clearly pointed to murder.

On Monday, 5 April 1982, Susan Barber and Richard Collins were arrested and charged with conspiracy to murder. Richard immediately said that Susan had told him that she wanted her husband dead, and had suggested that he should cut the hydraulic brake lines on Michael's car. Naturally he refused to do any such thing. A second piece of evidence he offered against Susan was in recalling how she had poured the remains of a bottle of Michael's medicine down the kitchen sink after the two of them had returned from visiting him in Hammersmith Hospital. During that visit she had been asked by medical staff whether it was possible that her husband might have taken poison.

Susan Barber told the police that she resented her husband finding her with Collins on the day of the ill-fated fishing trip and had resolved to end the marriage. She admitted that one evening she had put Gramoxone weedkiller into the gravy accompanying the steak and kidney pie she served for his evening meal, and had watched while he ate it. The dose was not large enough to be fatal, however, and only gave him a severe sore throat for which he obtained some medicine. In fact, husband Michael was to suffer an upset stomach and sore throat for several weeks, until Susan gave him a much larger dose.[vi] She told the police that she had come across the weedkiller in the garden shed, and it was there because Michael had once worked for a landscape gardening company. Moreover, the Gramoxone he had stolen from his former employers was manufactured during the period before ICI began adding warning chemicals to it.

In November 1982, Susan Barber and Richard Collins appeared at Chelmsford Crown Court before Mr Justice Woolf. Susan was charged with murder, conspiracy to murder, and of administering poison with intent to injure. Collins was charged only with

---

[vi] Paraquat is not a cumulative poison and it is more than likely that Susan was unaware of what constituted a fatal dose and kept increasing the amount she added to the gravy until she succeeded in killing him.

conspiracy to murder. Initially both pleaded not guilty. Susan Barber admitted putting the poison into her husband's food, but maintained that she didn't intend to kill him, she just wanted to make him ill enough for her to get away without him coming after her. However, the evidence was overwhelming, particularly that provided by the expert witness for the prosecution, Dr Bernard Hart, who had previously carried out extensive clinical researches into the effects of paraquat. The two eventually agreed to plead guilty. Collins, on whose behalf strong evidence of good character was provided, was sentenced to two years' imprisonment. Susan Barber pleaded that she had not intended to kill her husband but said she just wanted him to suffer as she had suffered, and admitted giving her husband two doses: "I mixed it with the gravy, and I gave him a second lot because the first did not seem to work."

In sentencing her to life imprisonment (minimum ten years) for the murder of Michael Barber, Mr Justice Woolf repeated what occurs to many people when murder by poisoning has taken place, and which judges down the centuries have quoted: "I cannot think of a more evil way of disposing of a human being." This was his way of expressing his revulsion at what Susan Barber had done, and no doubt all those in court would have agreed.

CHAPTER 10

Polonium and the Poisoning of Alexander Litvinenko

A word shown in **bold** *indicates that further information will be found in the Glossary*

10.1 A Novel Poison

Poisoning was the leading news story around the world during the autumn of 2006, when a former **KGB** officer, 43-year-old Alexander Litvinenko, became the first person ever to be deliberately poisoned by the element polonium—or at least the first we know about. On 23rd November he died in a London hospital from the intense radiation emitted by polonium-210, having been secretly given it in a cup of tea some three weeks earlier. What his assassins had not expected was that the poison would be identified, and that this would then allow detectives to deduce where it had originated, where it had been administered and by whom, and the places which the murderers had previously and subsequently visited.

It may never be revealed in what molecular form the polonium-210 was given to Litvinenko. Most likely it had been dissolved in water as polonium chloride. Unlike the other poisons discussed in this book, it is not a *chemical* property of the poison which is responsible for its toxicity; it is the damage caused by the alpha (α) particles emitted by polonium atoms as they convert to the element lead. Alpha particles as such are relatively harmless—a sheet of paper will stop them—but when they are ejected from an atom

Molecules of Murder: Criminal Molecules and Classic Cases
By John Emsley
© John Emsley, 2008

inside a living cell they can easily disrupt vital parts of its mechanism. Litvinenko's cells were bombarded with millions of such particles every second. Indeed, it was remarkable that he survived as long as he did, but in doing so he provided the evidence that identified both the poison and those responsible.

10.2 Polonium

Uranium ores contain about 100 µg of **polonium** per tonne, which is a mere 100 parts per trillion. Miniscule as this amount is, it can be separated, and this is exactly what Marie and Pierre Curie did in 1898 in an old shed in Paris. They were trying to discover what made the uranium mineral pitchblende[i] so much more radioactive than could be explained by its uranium content alone. The answer was a hitherto unknown element which they named after Marie's native country, Poland. The existence of polonium had been foreseen by Dmitri Mendeleyev in 1891 on the basis that there was a vacant slot in the periodic table below the element tellurium. He even predicted that it would have an atomic weight of 212. As we now know, polonium has several isotopes, ranging in mass from 192 to 218; the one the Curies isolated was the naturally occurring isotope of mass 209. All polonium isotopes are radioactive, but that isolated by the Curies has the longest **half-life**, 102 years. The half-life of the isotope polonium-210 which killed Litvinenko is 138 days.

Before the advent of nuclear reactors the only source of polonium was its tedious extraction from uranium ore, but that did not prevent it finding uses. In the early part of the twentieth century it was to be found in textile mills and in factories making photographic plates. It was designed to reduce static electricity, which in the mills caused unpleasant electric shocks to operatives, and in the case of photographic plates caused dust to be attracted to them. The polonium was electroplated on to the surface of metal foil and devices of this type were able to remove the static electricity charge, which is caused by a build-up of electrons, by virtue of the α-particles emitted. Alpha particles are in effect the nuclei of helium atoms. They carry a positive charge and each requires two electrons to form helium gas, which means that they are able to

---

[i] Pitchblende is principally uranium oxide, UO_2.

mop up any electrons they come across. The antistatic devices are still very much in use, and present no threat to humans because α-radiation is unable to travel more than a few centimetres before being neutralised by collision with molecules in the air, nor are they able to penetrate skin. Alpha particles are only dangerous when they are generated inside the human body.

No one now extracts polonium from uranium ores. Instead it is made in gram quantities by bombarding bismuth with neutrons in a nuclear reactor. Bismuth exists in only one isotopic form, bismuth-209, and when this is exposed to neutrons it absorbs one of them to produce the radioactive isotope bismuth-210. This has a half-life of five days, and when it decays it ejects a β-particle (an electron) from its nucleus and converts to polonium-210. The entire worldwide supply of polonium for commercial use is produced in Russia at the Ozersk nuclear reactor near the city of Chelyabinsk, east of the Ural Mountains. From there it goes to the Sarov nuclear facility near the city of Samara in southeast Russia, where the polonium is separated from the bismuth and sealed into capsules. Almost all the output is taken to St Petersburg to be exported by air to the USA for commercial distribution. It is not difficult to purchase polonium provided one has the necessary licence. United Nuclear Scientific Supplies of New Mexico is an approved supplier, but they only sell it in tiny amounts. (When the Litvinenko affair became news they assured visitors to their web site that a person would have to buy 15 000 of their polonium 'needles' to accumulate a toxic dose.)

Polonium has been used as an energy source in space exploration. A capsule of polonium containing only 1 g of the element is able to maintain a temperature of 500 °C due to the intense α-radiation, and such a capsule will generate 520 kJ of energy every hour. It provides an ideal lightweight energy supply for space satellites and was also used to generate thermoelectric power for the various rovers which Soviet Russia sent to explore the Moon during the years 1970–1974. A more sinister use of polonium has been in the triggers for atomic bombs. In combination with beryllium it can release a spray of neutrons with which to bombard uranium, thereby splitting its atoms and generating nuclear energy, and under the right conditions initiating an explosive chain reaction. Polonium used with beryllium can also be a source of neutrons for research purposes.

Polonium in the Human Frame

Polonium has no biological role, but trillions of its atoms are present in our bodies as polonium-209, which is also an alpha-emitter. These atoms arise from natural sources such as the food we eat (all soil contains traces of uranium, and this produces polonium) and the air we breathe (this contains traces of radon gas, which also decomposes to polonium). Polonium is even detectable in tobacco smoke. The amount of polonium excreted by the average person per day is one-tenth of a femtogram, which is 1×10^{-16} g, equivalent to 300 billion atoms. The biological half-life of the polonium present in the human body is about 50 days, which implies that only 1.5% of the total is excreted each day.

Despite the fact that the average person carries around 7×10^{-12} g (7 picograms) in their body, polonium is regarded as one of the deadliest substances known. One microgram (10^{-6} g) of polonium-210 given as a single dose is around a million times greater than the quantity normally present in the body; it contains around three quadrillion atoms (*i.e.* 3 000 000 000 000 000, or 3×10^{15}). Even if only 10% were absorbed into the bloodstream this would provide each living cell with scores of polonium atoms, every one of which could bond to proteins and remain there like tiny time bombs.

Radioactivity is measured in becquerels (Bq), a becquerel being the number of radioactive disintegrations occurring per second. The quantity of polonium-210 emitting 1 Bq is 5.9×10^{-15} g, which means that 1 μg of polonium-210 emits 170 million α-particles per second.

Polonium is not the only radioactive isotope in the human body. Indeed most of the radioactivity our body encounters arises from natural radioactive isotopes such as carbon-14, potassium-40 and uranium-238. Limited radiation damage is continuously repaired by the cells, but when millions of α-particles are being ejected every second the repair system becomes overwhelmed and death is certain.

Polonium was extensively researched during World War II as part of the Manhattan Project for making atomic bombs, and then more thoroughly in the 1950s and 1960s in nuclear energy projects in the USA. Tests on animals revealed how dangerous its α-emissions inside the body could be. This research was carried out by J Newell Stannard and George W Casarett at the University of Rochester and was reported in the official publication, *Radiation Research, Supplement 5: Alpha Particle Emitter, Polonium-210.* What this reveals is that the danger of ingesting a mere microgram of polonium-210 can be fatal. This quantity is about the size of a particle of dust and is barely visible.

Polonium is a restless element. It is quite volatile, and about half of a sample left open to the atmosphere will evaporate within three days. This can be avoided by encapsulating it, but anything less than a perfect seal means that some will escape to leave a trace behind, a fact that was to provide vital forensic evidence in the Litvinenko affair. Although in theory a microgram of polonium-210 should suffice to kill, it is most likely that a tenth of a milligram, which is 100 micrograms, was the dose given to Litvinenko. No medical treatment could have saved him, because there is no antidote for the radiation damage caused by α-particles. On the other hand, had it been realised within a few hours that he had been poisoned with polonium-210 it might have been possible to remove the unabsorbed polonium-210 remaining in his stomach and intestines, and some of the damage to his system could have been mitigated.

Litvinenko was not the first person to die of polonium poisoning. It was blamed for the accidental death of Irène Joliot-Curie, the daughter of Marie Curie. Irène died of leukaemia caused by exposure to polonium-209 after a sealed capsule burst on her laboratory bench. This longer-lived isotope took more than ten years to do its deadly work. We now know that polonium collects mainly in the liver, kidneys and bone marrow, and it would have been the last of these which caused her death.

Radiation detectors of the Geiger counter type are not suitable for detecting α-rays. What is required is a scintillation detector, which records the tiny flash of light emitted when an α-particle hits another atom. Zinc sulfide is the sensitive chemical present in scintillation counters, and while a single photon of light is difficult to detect, a photomultiplier can be used to enhance the effect. The energy of an α-particle is determined by the type of atom from which it originates. Those emitted by polonium-210 have a characteristic energy of 4.976 MeV (million electron volts). An electron volt is an incredibly small unit, but is more than a million times the energy needed to break a chemical bond, which explains why α-particles are so dangerous when generated within a living cell. To detect polonium-210 in a suspected location a probe is used which can detect the α-particles if it is less than 3 cm from a contaminated surface. An audible click is produced in the ratemeter attached to the probe.

When polonium is consumed it quickly causes vomiting. This removes the majority of the poison, but not before some has passed through the stomach into the intestines where it will begin to be absorbed. How quickly this takes place will depend on its chemical composition, but it was probably given to Litvinenko as polonium chloride ($PoCl_2$), which is soluble in water. Once it has passed through the gut wall and entered the blood stream the polonium will move around the body and enter the cells of various organs. Cells which have a high rate of division, such as hair follicles, are quickly affected, but wherever the polonium is present it will wreak havoc as the atoms disintegrate. The immune system will break down and there is a dramatic drop in white blood cells. The polonium-210 which passes into the liver damages the only organ which has the potential to remove this alien chemical from the body. The most obvious sign of liver damage is jaundice. Meanwhile the polonium-210 which remains in the intestines clings to the gut wall, leading to peritonitis and ultimately to toxic shock syndrome. The polonium-210 which lodges in the cells of the heart will cause irreparable damage to that vital organ. Litvinenko experienced all of these symptoms.

In the weeks following Litvinenko's death around 730 people who could have possibly been in contact with him, or who might have been at a contaminated site, were tested for exposure to polonium and 596 of them showed no sign of contamination, while 85 had slightly raised levels in their urine but not a cause for concern. Test results on another 35 people indicated definite contact with polonium-210, although not enough to affect their health, and there were 17 people whose urine samples showed significantly elevated levels but who were still not thought to be at long-term risk of cancer.

The test for polonium-210 in urine requires a volume of 1 litre to be collected over a period of a day. This is boiled with 200 ml of concentrated nitric acid and evaporated to dryness. The residue is redissolved in concentrated hydrochloric acid and again evaporated to dryness. The new residue is again redissolved, this time in a minimum of dilute hydrochloric acid, and the resulting solution will spontaneously deposit any polonium present on to a silver disc. The α-emissions from the disc are counted over a 12-hour period to assess any activity greater than 20 millibecquerel (mBq),[ii] indicating exposure above the expected background level. This method has in the past been used to measure the natural excretion rate of polonium from the diet and from smoking, and results are normally between 5 and 15 mBq per day. Some of those in the London investigation were found to have levels in excess of 30 units.

Surfaces exceeding 10 Bq per cm^2 were deemed to be contaminated and were cleaned with wet wipes which were then disposed of as low-level radioactive waste. Where the polonium was found on wood, to which it tends to cling, then the answer was simply to varnish over it. Similarly, contaminated walls were repainted. Taps on sinks tended to react chemically with the polonium, anchoring it firmly to the surface, and were replaced. At a meeting at the Royal Society of Chemistry on 13 July 2007 it was reported that 50 locations around London were eventually found to be contaminated with polonium-210 and were dealt with by contractors employed by the local authority in which the contamination was found. The clean-up was supervised by members of the Government Decontamination Service (GDS). It was revealed that in some cases, notably the bathrooms at the Millennium Hotel, the polonium had penetrated the porous surfaces of bathtubs and wash basins and could not be washed off; these were broken up with sledge hammers and the contaminated sections isolated for special disposal. It took the GDS almost three weeks to decontaminate the hotel.

Polonium-210 must have seemed the perfect poison for an assassination. Indeed, secret files from 1948 released by the US Government in 2007 showed that the US Army Services had considered using it as a weapon, referring to it as 'a new concept in warfare'. To what extent it was developed, and whether it was ever used as a means of assassination, is not revealed because the documents were heavily censored before declassification.

Polonium-210 must have seemed the ideal poison with which to assassinate someone. It emits no tell-tale gamma rays, which could

---

[ii] A mBq is a millibecquerel, *i.e.* one thousandth of a becquerel. An activity of 20 mBq over a period of 12 hours amounts to 864 disintegrations.

easily be detected by monitors at airports. Only a microscopic amount of the poison is needed—too tiny to be revealed by chemical analysis—and it can kill within days and death would be attributed to other causes, such as liver failure. As far as the murderers of Litvinenko were concerned, it did its job—but it turned out to be a less than perfect agent for disposing of people because once it was identified as the cause of death, it pointed an accusing finger straight at the murderer.

10.3   The Assassination of Alexander Litvinenko

Alexander Vaterovich Litvinenko was born in 1962 in the city of Voronezh, 500 km south of Moscow. (He was always known as Sasha.) He joined the army at 18 straight from high school and quickly rose to platoon commander. In 1988, when he was 26, he was selected to join the Russian Secret Service, the KGB as it was then known. (Today it is called the FSB, an abbreviation of the Russian words for Federal Security Service.) At first he served in the Division of Economic Affairs but was later transferred to the anti-terrorist group, where he was promoted to the rank of lieutenant colonel.

After the fall of the USSR in 1991 Litvinenko found himself checking on crime bosses in the seedy world that the post-communist privatisation free-for-all had produced. When he realised that leading members of the FSB were themselves in cahoots with criminal elements, he and some of his colleagues committed the unforgivable sin, at least as far as the secret service was concerned, of going public. They called a press conference in November 1998 and told the media what was going on. Thereafter Litvinenko became *persona non grata* with his employers—indeed he was arrested on trumped up charges and jailed for several months. These murky goings-on are recounted in the book, *Death of a Dissident*, by Alex Goldfarb and Marina Litvinenko, Sasha's widow. He had also been friendly with one of the new Russian billionaires, Boris Berezovsky, after he had uncovered a plot to kill the entrepreneur and thereby saved his life. Berezovsky was later to show his gratitude by helping Sasha find a new life in London.

Sasha, his wife Marina and their son Anatoly, went to Turkey in November 2000, ostensibly on holiday. They opted to return to Moscow *via* a connecting flight in London, saying they wanted to shop at Heathrow airport. Once in London, however, Sasha applied for political asylum and a temporary permit was granted to the family, made permanent six months later.[iii] Berezovsky had also deemed it prudent to leave Russia permanently once ex-KGB man Vladimir Putin became president and showed his determination to claw back the state assets that his predecessor, Yeltsin, had sold off too cheaply. Berezovsky settled in London and employed Litvinenko, paying him a salary of £5000 a month. In February 2003 Berezovsky even bought a new three-bedroomed house in Muswell Hill, which he rented to the Litvinenkos, and they moved out of their rented flat in Kensington.

Sasha appears to have been employed by Berezovsky to collect anti-Putin evidence and ensure it was published. This culminated in his controversial book, *Blowing up Russia*, parts of which were serialised by the publication *Novaya Gazeta* and turned into a television documentary, financed by Berezovsky. Though this was banned from being broadcast in Russia, it circulated widely on video. The book accused the FSB of a series of bombings of apartment blocks in 1999, claimed at the time to be the work of Chechen rebels. On 9 September the first of the explosions occurred in Moscow, killing 94 occupants, to be repeated on 13 September when a further 119 died. Three days later a similar attack in the city of Volgodonsk killed 17 people, and on 23 September yet another explosion might have taken place had a fourth device exploded in the basement of a block of flats in the town of Ryaza—this was discovered by the police before it could be detonated. The FSB later claiming they had placed it there as part of a training exercise. Whether there is any truth to Sasha's claims we will never know for certain, but the explosions provided the Kremlin with the necessary justification for launching attacks on this breakaway republic.

Sasha was reputed to be involved in all manner of other activities, but it is now virtually impossible to untangle the web of intrigue which surrounded him. He was supposed to have removed secret files from the FSB, and there were hints that he had been working for British Intelligence, MI6. Fascinating as his life was, our main interest is to understand how he was poisoned. Clearly there were those who wanted this troublesome individual out of the way and

---

[iii] Because political refugees are at continued risk they are allowed to choose a new name, and officially Sasha became Edwin Redwald Carter. This did not prevent the Russians tracking him down by searching Marina's mother's luggage when she returned from a visit to London—among her belongings they came across the family's new address.

they hatched plot to kill him in the autumn of 2006, using what they thought would be an undetectable poison: polonium-210.

On the day of his poisoning, Wednesday, 1 November 2006, Sasha caught the 134 bus into central London using his Oyster card. This is a prepaid season ticket which deducts money for every journey that its holder makes; the card is passed over readers which register the beginning and end of a journey. This identified the actual bus he travelled on, and it was later checked and shown to be free of polonium-210. The bus arrived in central London at 11:30 am, and Litvinenko then went into a shop and bought a newspaper and bottle of water. What he did for the next three hours is unclear, although his movements will almost certainly have been recorded on some of the tens of thousands of CCTV cameras which are located in London streets, on public transport and inside buildings. His presence in the shop was filmed, but if his subsequent movements were also caught on camera the police are keeping quiet about them.

Sasha next surfaced at 3 pm at the Itsu sushi bar on Piccadilly, a venue later discovered to be heavily contaminated with polonium-210. There he met Mario Scaramella, ostensibly an Italian academic, but actually a member of an official investigation set up by the Italian Government to look for Russian spies in Italy. (Scaramella also appears to have been involved in some of the murkier goings-on within Italian politics.) It is believed that Scaramella handed documents to Sasha naming the killers of the Russian investigative journalist Anna Politkovskaya, shot in the head on 7 October at her apartment block in Moscow. Sasha spoke at a commemoration ceremony at the Frontline Club in London, during which he openly accused President Putin of ordering her murder.[iv]

Scaramella was arrested soon after Litvinenko's death and his urine tested positive for polonium-210, but this was only at the expected background level and it was clear that he was not the assassin. It was at 4:30 pm at his next meeting place, the Pine Bar of the Millennium Hotel in Grosvenor Square, that Sasha came face

---

[iv] On 28 August 2007 Russia's chief prosecutor, Yuri Chaika, announced that 10 people had been arrested in connection with her murder, which had been carried out by a hit man who shot her three times in the lift of her apartment block. He had been caught on camera but was wearing a white baseball cap which made identification impossible. The men arrested were described as rogue elements within the FSB but paid for by enemies of the Kremlin based abroad.

to face with his murderer. There he met three Russians, newly arrived in London and who had come to watch the key football match between Arsenal and CSKA Moscow.[v] They were Andrei Lugovoy, Dmitry Kovtun and Vladislav Sokolenko, and one of them put polonium-210 in a cup of tea which Sasha drank. He later said there had been a pot of green tea waiting for him when he arrived at the meeting. After he had consumed the fatal drink he contaminated everything he came into contact with—as also did his murderer.

After the meeting Sasha went to Berezovsky's office in Down Street, where he met Akhmed Sakayev, who drove him back to Muswell Hill in his Mercedes. Sakayev was a leading member of the Chechen exiles in London. The car was later impounded on account of its high level of polonium-210 contamination. The fax machine at Berezovsky's office, which was used by Sasha, was also found to be contaminated.

Sasha went to bed at 11 pm but by midnight he was vomiting, and this continued throughout the night. On the morning of 2 November he rang Lugovoy and Kovtun at the Millennium Hotel to call off a planned meeting arranged for 10 am, saying he was too ill. The following day the two flew back to Moscow on flight BA847. Sasha remained in bed but his condition steadily worsened, and during the following night he became so ill that Marina rang for an ambulance. He was taken to Barnet General Hospital, and the Accident and Emergency Unit diagnosed a stomach infection. He was sent home again and advised to take painkillers, drink plenty of water and rest until he was better. When his own doctor saw him the following day he too diagnosed a stomach infection, but of such a virulent nature that he clearly need hospital treatment. He immediately ordered an ambulance and sent him back to Barnet General Hospital, where he was given powerful antibiotics but there was no improvement in his condition. The doctors noticed that his white blood cell count was very low, indicating that his bone marrow had stopped producing the new cells needed to fight infection.

By Monday, 6 November, Sasha could barely talk and was being fed by tube as he could no longer swallow. His skin had turned yellow, indicating liver dysfunction, and he was tested for the two most likely causes, hepatitis and AIDS, but neither was the cause.

---

[v] This ended in a 0–0 draw.

Then his hair began to fall out. By now Sasha realised that he had been poisoned, and he agreed to give an interview for the BBC Russian Service in which he claimed that this had happened at the sushi bar.

Sasha continued to deteriorate and on 17 November was transferred to University College Hospital in central London, one of Britain's leading hospitals. By now he was completely bald. Specialists assumed the cause to be thallium, a favourite poison used by secret agents.[vi] It had been used by a serial poisoner in the 1970s, Graham Young, to kill his colleagues at work. They had died undiagnosed in London hospitals, but in any case there was at the time no known antidote. But specialists in London were now skilled in recognising, diagnosing, and eventually treating thallium poisoning, skills they put to use in the 1980s, as we shall see. The antidote for thallium poisoning is **Prussian Blue**. This was given to Sasha, but it brought no relief and the treatment was discontinued after three days when analysis of his blood showed no evidence of thallium poisoning.

His thallium level was in fact three times greater than the background level, but in itself this was still far too low to cause the symptoms of thallium poisoning. It was therefore assumed that he had been poisoned with the *radioactive* thallium-201, a short-lived isotope with a half-life of only a few days, meaning that it would have almost disappeared by the time his blood was analysed. The doctors treating him finally concluded that his condition could only be explained as some form of radiation poisoning, yet tests for gamma rays using a Geiger counter were negative and thallium-201, a gamma emitter, and was therefore ruled out. Samples of Sasha's urine were sent for analysis to the Atomic Weapons Research Establishment at Aldermaston, Berkshire, where it was established that Sasha's radioactive condition was not due to thallium-201, but to the deadlier polonium-210.

On 20 November Sasha was moved into intensive care, connected to a series of drips and propped up on pillows. Once the media became aware of what had happened a photograph taken that day was flashed around the world. On 22 November he deteriorated rapidly and during the early hours of 23 November his

---

[vi] My previous book *Elements of Murder* had chapters on thallium, including its use by secret agents as a poison, and I was interviewed for television to explain how this poison works and why it causes the hair to fall out.

heart stopped. It was restarted, and it was later that day that his doctors finally learned what was killing him. The analysis of his urine had shown dangerous levels of polonium-210, but the doctors knew that there was no antidote. It hardly mattered, because at 9:21 pm that evening he died.

Things now began to move rapidly as the authorities sprang into action. Professor Patricia (Pat) Troop, chief executive of the Health Protection Agency (HPA)[vii] had already been alerted by Dr Roger Cox, director of the HPA Centre for Radiation and Environmental Hazards at Chilton, Oxfordshire, and scientists from the centre arrived at the hospital with orders to screen everyone who had been in contact with Sasha and examine all the areas where he had been. Cox had telephoned Troop the evening of the death to tell her what they had found. She was in Helsinki at the time but flew back to the UK on the first flight the following morning. A dangerous radioactive isotope had been brought into London and was contaminating parts of the city. Scientists from the HPA and the Atomic Weapons Establishment now began to investigate every conceivable location which Sasha might have visited on 1 November, the day when he had quite clearly been poisoned with many times the fatal dose.

Within hours of his death a document purporting to be his final testament was released to the media, in which he accused the Russian president of killing him, and so began a campaign to blame Putin and the FSB for his murder. Sasha dictated the statement, thanking his wife for her love, his doctors and nurses for their care, and the British Government for granting him political asylum. The rest of his deathbed statement accuses Putin of being unfit for office and of instigating his murder in order to silence him.

Alexander Litvinenko was buried at Highgate Cemetery in North London on Thursday, 7 December 2006, but not in a lead-lined coffin as reported in the media. His most affected organs had been removed at autopsy for disposal elsewhere. There is a certain irony in the fact that this cemetery is also the final resting place of Karl Marx. What was more ironic, and a shock to those who attended his funeral, was that a mullah presided over the graveside ceremony, claiming that Sasha had converted to Islam the day before he died. That conversion had been arranged by Sakayev and

---

[vii] The HPA was set up in 2003 to protect the health and wellbeing of the public, including preparation for terrorist attacks in the form of chemical, biological or radioactive materials.

performed by a mullah who led Sasha through the appropriate prayer which made possible his deathbed conversion. This is not as improbable as it might seem, because Sasha's writings showed his close sympathy with the Chechen people, who are mainly Muslims, and whom he believed to be the victims of a war initiated by the Kremlin with the help of the FSB.

So who murdered Sasha Litvinenko? The poison itself is an important clue. Few, if any, organisations have the facilities to produce it and concentrate it to the form suitable for an assassination, and in a way that did not expose the assassin to risk. It is almost certain that the polonium-210 originated from the infamous poisons laboratory in Moscow known as the Kamera (the chamber), said to be run by Department 12 of the SVR, the Russian Foreign Intelligence Service. Whatever its source, it ended up in London in the locations where Sasha had been, and indeed where those he had met on that fateful Wednesday had been. Sasha's trail of polonium-210 included the Pine Bar in the Millennium Hotel, Berezovsky's office in Down Street, the Mercedes which had taken him back to Muswell Hill, and of course his home and the two hospital wards where he was treated. The heaviest contamination of polonium-210 was found in the kitchens of the Millennium Hotel, where seven staff tested positive. It was also present at the Itsu sushi bar in Piccadilly where Sasha had met Scaramella, and which had initially made him the prime suspect.

Polonium-210 was also found in locations where Sasha had definitely not been, such as two British Airways 767 planes operating on the service between London and Moscow. The Russians who met Sasha had also travelled to London on the Russian airline, Transaero, but rather unsportingly that airline refused permission for its seats to be tested for contamination. Polonium was also found at the Emirates stadium in Arsenal on the seats occupied by the Russians Sacha had met at the Millennium Hotel.

There could well have been more than one dose of polonium-210 brought to London, there to be combined to make up the fatal dose. It transpired that Sasha had previously met Lugovoy and Kovtum on 16 October, and again later that month. These two men now became the prime suspects, and travel records confirmed that they had travelled to London a number of times during the months leading up to the fatal meeting. On 16 October they had stayed at the Parkes Hotel in Knightsbridge, near Harrods, and inspection of the rooms they occupied showed them to be contaminated with polonium-210, indicating that some of the poison was already in the UK at that time. That was a brief visit, and Lugovoy returned to Moscow but came back again on 25 October, this time by British Airways. Their records showed exactly which seat Lugovoy had occupied, and this too was found to be emitting the alpha rays characteristic of polonium-210, as also was the overhead locker, suggesting that his briefcase also was radioactive. This time he stayed at the Sheraton Hotel in Park Lane, and once again his room, 848 on the eighth floor, was contaminated, and at higher levels that those at the Pine Bar. Then on 28 October he flew back to Moscow on the same British Airways plane, but contaminated a different seat. That same day his confederate Dmitry Kovtum flew from Moscow to Hamburg to see his German ex-wife and their child, but he flew on the Russian airline Aeroflot, to which plane the detectives were denied access. However, the flat where his ex-wife lived was found to be contaminated, as was the car park outside, so clearly Kovtum had been carrying some of the poison with him.

On 31 October both Lugovoy and Kovtum flew to London once more by British Airways and again the seats were contaminated with polonium-210. These various discoveries suggest that portions of the polonium were indeed brought to London on three separate occasions, and they had carried it with them while awaiting an opportunity to put it into Sasha's drink. On the evening before they met him, Lugovoy went to see his old friend Berezovsky at his Down Street office—he had worked for the billionaire in the 1990s as his bodyguard—and the leather sofa on which he sat was later shown to be contaminated. Afterwards he went to the Millennium Hotel, where he was staying with Kovtum and Sokolenko. The rooms used by Lugovoy and Kovtum, 101 and 441, were later found to be heavily contaminated, but not the one occupied by Sokolenko.

The radioactivity at the sushi bar was now explained because Sasha had met Lugovoy and Kovtum there on 16 October, and the polonium contamination had come from one or both of these individuals. Indeed, when the CCTV pictures from that meeting were scanned it turned out that they were all sitting at the table which was radioactive, whereas the table Sasha used for his meeting with Scaramella on 1 November was not. On 5 January 2007 detectives found polonium-210 at the Pescatori restaurant on Piccadilly, also visited by Lugovoy, and at the offices of two security firms, Erinys and Titon International, to which he had been.

Deputy Assistant Commissioner Peter Clarke of Scotland Yard's Anti-Terrorist Command was put in charge of investigating Litvinenko's death, and eventually named Lugovoy as the person who should stand trial for his murder. Lugovoy was a former KGB officer, and Kovtum was his business partner; they had known each other since their youth. Lugovoy had worked for a time as a Kremlin bodyguard and then became head of security at ORT, a television channel controlled by Berezovsky. It was during this time that Lugovoy met Sasha. Lugovoy is now a multimillionaire and runs a security business, providing bodyguards for a number of Moscow's other millionaires.

When London detectives went to Moscow to interview Lugovoy and Kovtum, these two protested their innocence and claimed that they too had been poisoned. Kovtum appeared to have lost all his hair and both were admitted to hospital. However, when the Scotland Yard team interviewed them only a few days later they appeared to be in good health. The Russian Prosecutor General's Office even began its own investigation into their claims that they had been the victims of attempted murder. A wide variety of theories were advanced by the Russians as to what might have taken place in London. One was that Litvinenko and Berezovsky were working on a dirty bomb and that they had contaminated the two visiting Russians. Another was that Berezovsky himself had organised the assassination of his employee Sasha, because he had become a loose canon and his wild accusations were counterproductive to Berezovsky's campaign against Putin. Sasha was even said to be accusing Putin of being a paedophile. By the time he was murdered, Berezovsky had cut Sasha's allowance to £1500 a month, but by that point he was earning most of his income as a security consultant.

In April 2007 Russian investigators came to the UK to question Sakayev and Berezovsky about Sasha's death, no doubt hoping to lay similar charges against them as the Crown Prosecution Service in London had done against Lugovoy. On 28 May 2007 the Russian authorities refused an extradition request for Lugovoy, saying this was contrary to Russian law. However, they hinted that they might be prepared to exchange Lugovoy for Berezovsky. On 9 July, the UK was formally told that Lugovoy would not be extradited.

In August 2007 Lugovoy spoke to journalists in London *via* a video link and told them that MI6 had been involved in the murder of Litvinenko, and that detectives should focus their attention on Berezovsky. He even went so far as to suggest that the polonium-210 had been deliberately placed in locations to incriminate him, and then only after he had left London. Lugovoy entered politics in the autumn of 2007 and became a key member of the LDPR, the Liberal Democratic Party of Russia. In the general election held on 2 December he won a seat in the State Duma, and as such acquired immunity from prosecution.

In his well-researched book, *The Litvinenko File*, Martin Sixsmith concludes that Sasha's poisoning was organised by a group within the FSB working independently of the Kremlin, and that this was a revenge killing for betraying his former comrades. In fact, Litvinenko had been warned as long ago as 2000 that he was to be killed because he was a traitor. In his book, *Death of a Dissident*, Alex Goldfarb also believes Litvinenko's murder had nothing to do with Putin, simply because he was such a minor player and hardly worthy of the President's consideration. On the other hand, his death might have been part of a plot to discredit Berezovsky.

Whoever it was who wanted Sasha dead no doubt thought that polonium-210 was the perfect poison. Indeed it might have been, had Litvinenko not lingered for so long that the medical authorities had time to learn that he was suffering from radiation poisoning. A more massive dose of polonium-210 could have killed him earlier and his death ascribed to a virulent virus. A smaller dose might have weakened his immune system and vital organs, so that he would have succumbed to what appeared to be natural causes a few months later. But the assassins got it wrong, and it seems most unlikely that this agent will be used again. Any suspicious death such as Litvinenko's would include tests for α-radiation, and the use of polonium-210 would immediately be revealed and the assassins identified.

10.4   Assassination by Poison Down the Ages

Sasha Litvinenko is the latest in a long series of high profile victims of political assassinations involving poison. Indeed, this nefarious activity has been going on for more than 2500 years, probably even longer. Another quite recent politically motivated poisoning was that of Georgi Markov, a story recounted in Chapter 1.

No doubt by accident, early humans discovered that certain plants were poisonous, fatally so in some cases. Hemlock was such a plant, and it was a concoction made from this which the Greek philosopher Socrates drank when he was forced to commit suicide in 399 BC. Humans also discovered that certain minerals were dangerous, including orpiment and realgar, which are sulfides of arsenic, and mispickel, which is iron arsenic sulfide. (Orpiment is As_2S_3, realgar As_2S_2 and mispickel, FeAsS.) These became even deadlier on heating, when they were converted to **arsenic** (III) oxide (also known as white arsenic, As_2O_3). Unlike the sulfides, this dissolves in water to form a colourless, tasteless solution; a quarter of a gram (250 mg) of arsenic trioxide in a glass of wine can kill.

One of the more accomplished poisoners in ancient Rome was the infamous Agrippina. She probably used arsenic to murder her husband. That left her free to marry her uncle, Emperor Claudius, although there was the small matter of disposing of Claudius's wife, Valeria, but this was soon accomplished. Agrippina and the emperor were married shortly afterwards and she then prevailed upon him to marry his daughter Octavia to her son Nero. All that remained was to poison the emperor's natural son and heir apparent, Britannicus, and persuade Claudius to name Nero his successor. As soon as he did this he effectively sealed his own fate. She poisoned Claudius in 54 AD, and Nero became emperor at the tender age of 16. Ungrateful son that he was, five years afterwards he repaid his mother by having her stabbed to death. That same century also saw another Roman emperor, Titus, who reigned from AD 79–81, poisoned by his physician Valens acting on behalf of Titus's brother, Domitian. He succeeded Titus and reigned for 15 years, becoming notorious for his brutality.

The evidence for politically motivated poisonings of long ago is in general anecdotal, but occasionally there is forensic proof. Analysis of the remains of Pope Clement II, who died in 1047 AD, showed that he had been poisoned using lead. His remains lay undisturbed in a stone coffin in Bamberg, Germany, until in 1959 samples of his bones were analysed and showed a high level of lead in a pattern suggesting that he had had a high intake of lead during the final days of his life, probably added to the German wine of which he was particularly fond. In those days, and indeed up to the early 1800s, wine was preserved and sweetened with small amounts of litharge (lead oxide) to neutralise any acids present. This formed the sweet-tasting lead acetate (also known as sugar of lead). Adding more litharge than necessary would turn the wine into a decidedly unhealthy potion, indeed down the centuries whole communities were victims of over-leaded wine and many people died as a result.

Who would want to poison a Pope? In 1046 the previous Pope Benedict IX had been ousted on account of his dissolute lifestyle, and Clement II had been enthroned in his place by the Holy Roman Emperor, Henry III. Rumour has it that Benedict IX's agents poisoned his successor, and Benedict now appeared in Rome to reclaim the Papacy. His triumph was short-lived, however. In July 1048 Henry III again expelled him and a new Pope, Damascus II, was crowned.

The use of poison in the furtherance of political ends reached a fine art in Italy in the 1500s. The most notorious practitioners were Cesare Borgia and his sister Lucrezia, whose name is still synonymous with political poisoning. She devised a toxic concoction based on arsenic, and with it they murdered opponents of their father, who had become Pope Alexander VI. Some even think they went a step too far, because the Pope himself suddenly died in 1503 after attending a banquet with his son Cesare. His death was apparently due to drinking wine intended for someone else.

During the twentieth century the preferred poison for disposing of political opponents has been thallium sulfate. This was used by the KGB in the 1950s, and appeared to be an ideal poison since it takes two or three days before any symptoms appear. Occasionally it failed to kill, however, and this happened in 1957 to the former KGB agent, 35-year-old Dr Nikolai Khokhlov. (At one time he had himself been sent to Germany to poison an anti-communist émigré.) Khokhlov eventually defected to the West and he too became targeted. At a meeting in Frankfurt in September 1957 he was handed a cup of coffee laced with thallium sulfate, but he drank only a mouthful, and this saved his life. Nevertheless the thallium made him very ill, and eventually his hair fell out. Slowly he recovered, and is still alive and living in California.

Saddam Hussein's agents also used thallium sulfate for more than 20 years. In 1977 the religious scholar Mohsen Shubbar was given it, but survived. Three years later Salwa Bahrani, a prominent Shia, had his yoghurt poisoned and died a long and painful death. Another important Shia, Majidi Jehad, was given it in orange juice at a police station in Baghdad while collecting an exit visa to visit Britain. Soon after his arrival in London he was taken ill and admitted to hospital, where he died.

Throughout the 1980s dissident Iraqi scientists and clerics were similarly dispatched, including several living abroad, for example 44-year-old Abdullah Ali, who was resident in the UK. He became ill after visiting a restaurant in Notting Hill Gate, London, along with three men who had arrived from Baghdad. He died 15 days later, on 16 January 1988, at St Stephen's Hospital, Fulham. In his death-bed statement he said he suspected that his vodka had been spiked with the poison. The Iraqi agent in London responsible was reputed to be an attractive woman who had targeted members of the Patriotic Union of Kurdistan, a militant anti-Saddam Hussein group. She was alleged to have killed several key figures, including 38-year-old Sami Shorash and 40-year-old Mustafa Mahmoud.

In 1992 two high-ranking Iraqi army officers, Abdallah Abdelatif and Abdel al-Masdiwi, fell from favour and became ill. They escaped to Damascus, where they were granted emergency visas by the British Foreign Office and flown to London. Thallium poisoning was diagnosed and successfully treated. A similar occurrence took place with 31-year-old Kurdish resistance fighter Safa al-Battat, when he suspected that his drink of Coca Cola had been poisoned. He escaped to Britain, and his life too was saved.

Of course poisons can have effects which are not fatal, but which may damage the victim's aspirations. This is undoubtedly what happened to Viktor Yushchenko. In September 2004, this handsome 50-year-old Ukrainian politician was vying to become the next president of his country in a bitterly fought contest. He was poisoned with **tetrachlorodibenzodioxin** (TCDD), which turned his face into a mass of pustules and left him permanently scarred. TCDD is the most unpleasant of the dioxin group of chemicals and causes the skin disease, chloracne. Soon after he had been poisoned, Yushchenko was diagnosed with a pancreatic disorder and flown to a specialist clinic in Vienna. As a result of seeing him on television, a Dutch toxicologist, Bram Brouwer, suspected he was suffering from chloracne and requested a sample of Yushchenko's blood, which revealed a high level of TCDD. Yushchenko blamed the SBU (Security Service of the Ukraine), who had taken this action because he was opposing the Government's official candidate for the presidency.

So what secret poisons will assassins choose in the twenty-first century? Thankfully their choice is limited—indeed there is nothing they can now use which could escape detection. Advances in forensic analysis make this a certainty whenever a death is regarded as suspicious. Chemical poisons are easy to detect. Biological toxins such as ricin are more elusive, but the antibodies produced by the victim's body give the game away. Polonium-210 may have seemed likely to avoid these pitfalls, but as we now know it not only provides clues as to the cause of death, it enables the investigators to locate where it was given, identify the assassin, and trace it back to its country of origin. Maybe the days of political assassinations by poisoning are finally over. Maybe.

Further Reading

Web Sites

The home page of the US Department of Health and Human Services (www.atsdr.cdc.gov) has lots of information.
The home page of the UK Forensic Science Service (www.forensic.gov.uk) also has lots of information.

General References

Bennett, H. (ed.), *Concise Chemical and Technical Dictionary*, 3rd edn, Edward Arnold, New York, 1974.

Budavari S. (ed.), *The Merck Index*, 13th edn, Merck & Co Inc, Rahway, NJ, 2001.

Butler, I., *Murderer's England*, Robert Hale, London, 1973.

Camps, F.E. (ed.), *Gradwohl's Legal Medicine*, 2nd edn, John Wright & Son Ltd, Bristol, 1968.

Clayden, J., Greeves, N., Warren, S. and Wothers, P., *Organic Chemistry*, Oxford University Press, Oxford, 2001.

Cooper, P., *Poisoning by Drugs and Chemicals, Plants and Animals*, 3rd edn, Alchemist Publications, London, 1974.

Duffus, J.H. and Worth H.G.J. (eds.), *Fundamental Toxicology for Chemists*, Royal Society of Chemistry, London, 1996.

Emsley, J., *The Elements,* 3rd edn, Oxford University Press, Oxford, 1995.

Emsley, J., *Nature's Building Blocks,* Oxford University Press, Oxford, 2001.

Emsley, J., *Elements of Murder,* Oxford University Press, Oxford, 2005.

Evans, C., *The Casebook of Forensic Detection,* John Wiley & Sons Inc, New York, 1996.

Gerber, S.M. and Saferstein, R. (eds.), *More Chemistry and Crime*, American Chemical Society, Washington DC, 1997.

Glaister, J., *The Power of Poison,* Christopher Johnson, London, 1954.

Greenwood, N.N. and Earnshaw A., *Chemistry of the Elements*, 2nd edn, Butterworth Heinemann, Oxford, 1997.

Gunther, R.T. (ed.), *The Greek Herbal of Dioscorides*, trans. John Gooyer, Oxford University Press, Oxford, 1934.

Hawley, G.G., *The Condensed Chemical Dictionary*, Van Nostrand Reinhold, New York, 1981.

Hunter, D., *Diseases of Occupations*, 5th edn, Hodder and Stoughton, London, 1976.

Jacobs, M.B., *The Analytical Chemistry of Industrial Poisons, Hazards, and Solvents*, 2nd edn, Interscience, New York, 1949.

Johll, M.E., *Investigating Chemistry: A Forensic Science Perspective*, Freeman & Co, New York, 2007.

Julien, R.M., *A Primer of Drug Action*, 10th edn, Freeman & Co, New York, 2004.

Kaye, B.H., *Science and the Detective*, Wiley–VCH, Weinheim, 1995.

Kelleher M. and Kelleher C.L., *Murder Most Rare: The Female Serial Killer*, Dell Publishing, New York, 1998.

Kind S., *The Sceptical Witness*, Hodology Ltd and The Forensic Science Society, Harrogate, 1999.

Klaassen, C.D. (ed.), *Casarett & Doull's Toxicology: the Basic Science of Poisons*, 6th edn, McGraw Hill, New York, 2001.

Lewis, R.J. Sr, *Carcinogenically Active Chemicals*, Van Nostrand Reinhold, New York, 1991.

Lund W. (ed.), *The Pharmaceutical Codex,* 12th edn, The Pharmaceutical Press, London, 1994.

Macinnis, P., *Poisons from Hemlock to Botox and the Killer Bean of Calabar*, Arcade Publishing, New York, 2004.

McLaughlin, T., *The Coward's Weapon*, Robert Hale, London, 1980.

Mann J., *et al.*, *Natural Products*, Addison Wesley Longman Ltd, Harlow, UK, 1996.

Mann, J., *Murder, Magic and Medicine*, revised edn, Oxford University Press, Oxford, 2000.

Montgomery Hyde, H., *Crime has its Heroes*, Constable, London, 1976.

Ottoboni, M.A., *The Dose Makes the Poison*, 2nd edn, Van Nostrand Reinhold, New York, 1991.

Overton. C.E., *Studies of Narcosis*, Chapman and Hall, London, 1991.

Pasternak, C.A., *The Molecules Within Us*, Plenum Trade, New York, 1998.

Paul, P., *Murder, Under the Microscope*, Macdonald, London, 1990.

Pearce, J. (ed.), *Gradner's Chemical Synonyms and Trade Names*, 9th edn, Gower Technical Press, Aldershot, UK, 1987.

Polson, C.J. and Tattersall, R.N., *Clinical Toxicology*, EUP, London, 1965.

Rentoul, E. and Smith H., *Glaister's Medical Jurisprudence and Toxicology*, 13th edn, Churchill, Edinburgh, 1973,

Rowland, R., *Poisoner in the Dock*, Arco, London, 1960.

Schwarcz, J., *The Genie in the Bottle*, W. H. Freeman, New York, 2002.

Sharp, W.A. (ed.), *The Penguin Dictionary of Chemistry*, 3rd edn, Penguin, London, 2003.

Simpson, K. (ed.), *Taylor's Principles and Practice of Medical Jurisprudence*, 12th edn, vol. 2, Churchill, London, 1965.

Stevens, S.D. and Klarner, A., *Deadly Doses: A Writer's Guide to Poisons*, Writer's Digest Books, Cincinnati, Ohio, 1990.

Stolman, A. and Stewart, C.P., The absorption, distribution, and excretion of poisons, *Progress in Chemical Toxicology*, 1965, **2**, 141.

Stone, T. and Darlington, G., *Pills, Potions and Poisons*, Oxford University Press, Oxford, 2000.

Sunshine, I. (ed.), *Handbook of Analytical Toxicology*, Chemical Rubber Co, Cleveland, Ohio, 1969.

Thompson, C.J.S., *Poisons and Poisoners*, Harold Shaylor, London, 1931.

Thorwald, J., *Proof of Poison*, Thames & Hudson, London, 1966.

Timbrell, J., *Introduction to Toxicology*, Taylor & Francis, London, 1989.

Timbrell, J., *The Poison Paradox: Chemicals as Friends and Foes*, Oxford University Press, Oxford, 2005.

Trestrail, J.H. III, *Criminal Poisoning*, 2nd edn, Humana Press, Totowa, New Jersey, 2007.

Wade, A. (ed.), *Martindale: The Extra Pharmacopoeia*, 27th edn, The Pharmaceutical Press, London, 1977.

Watson, K., *Poisoned Lives*, Hambledon, London, 2004.

Weatherall, M., *In Search of a Cure*, Oxford University Press, Oxford, 1990.

White, P.C. (ed.), *Crime Scene to Court*, Royal Society of Chemistry, Cambridge, 2004. [This excellent and comprehensive book contains a chapter by R. Anderson devoted entirely to forensic toxicology.]

Wilson, C. and Pitman, P., *Encyclopaedia of Murder*, Arthur Barker, London, 1961.

Winger, G., Hofman, F.G. and Woods, J.H., *A Handbook on Drug and Alcohol Abuse*, Oxford University Press, New York, 1992.

Witthaus, R.A., *Manual of Toxicology*, William Wood, New York, 1911.

Wootton, A.C., *Chronicles of Pharmacy*, Milford House, Boston, 1910 (republished 1971.)

Ricin

Ball, P., *Murder under the Microscope*, Macdonald, London, 1990. [Chapter 14, Poisoned Pellets]

Schwarcz, J., *Let Them Eat Flax*, ECW Press, Toronto, 2005.

Hyoscine

Busia K. and Heckles, F., Jimsonweed: history, perceptions, traditional uses, and potential therapeutic benefits of the genus *Datura*, *HerbalGram*, 2006, **69**, 40–50.

Kotsias, B.A., Scopolamine and the murder of King Hamlet, *Arch. Otolaryngology*, 2002, **128**, 847.

Atropine

Holzman, R.S., The legacy of Atropos, the fate who cut the thread of life, *Anaesthesiology*, 1998, **89**, 241.

Lee, M.R., Solanaceae IV: *Atropa belladonna*, Deadly Nightshade, *Journal of the Royal College of Physicians of Edinburgh*, 2007, **37**, 77–84.

Morphine and Diamorphine

Cullen, P.V., *A Stranger in the Blood*, Elliott & Thompson, London, 2006. [The John Bodkin Adams case]

Walker, A., The pain and pleasure principle, *Chemistry in Britain*, 2002 (September), 24.

Adrenaline

Phelps, M.W., *Perfect Poison: A Female Serial Killer's Deadly Medicine*, Pinnacle Books, New York, 2003.

Chloroform

Farrell, M., *Br. Med. J.*, 1994, **309**, 1720.
Stratmann, L., *Chloroform: The Quest for Oblivion*, Sutton Publishing, Stroud, UK, 2003.

Carbon Monoxide

Foresti, R., *et al., Br. J. Pharmacol*, 2004, **142**, 453. [CORM-3]
Watson, A.A., *Forensic Medicine, a Handbook for Professionals*, Gower, Aldershot, UK, 1989.

Cyanide

Hunter, D., *Diseases of Occupations*, 5th edn, Hodder and Stoughton, London, 1976.
Melamed, A., *Chem. Rev.*, 1998 (September), 2. [Konzo]

Paraquat

Paul, P., *Murder, Under the Microscope*, Macdonald, London, 1990. [Chapter 23, Paraquat Pie]
Tullett, T., *Clues to Murder*, The Bodley Head, London, 1986. [Chapter 22, Poison Pie]

Polonium-210

Goldfarb A. and Litvinenko, M., *Death of a Dissident*, Simon & Schuster, London, 2007.
Sixsmith, M., *The Litvinenko File*, Macmillan, London, 2007.

Glossary

A word shown in **bold italics** *indicates another entry in the
Glossary under that name.*

Acetic acid – see **ethanoic acid**.

Acetylcholine (ACh) is one of the main neurotransmitters. When a
signal reaches the end of a nerve fibre it opens a channel which
allows calcium ions to move into cells, and these trigger the
release of ACh. This moves across the synaptic gap and locks on
to receptors on the membrane of the organ to be activated. After
it has completed its task the ACh is deactivated by an enzyme.

Adenine (A) is one of the bases present in DNA and RNA. The
others are guanine (G), cytosine (C) and thymine (T).

Adiponitrile is also known as 1,4-dicyanobutane and has the formula $NC(CH_2)_4CN$. This is converted to 1,6-diaminohexane,
$H_2N(CH_2)_6NH_2$, a precursor of **nylon**.

Adrenaline – see **epinephrine**.

Amygdalin consists of two linked glucose units attached to benzyl
cyanide. It occurs naturally in apple pips.

Anticholinergic chemicals block the action of *acetylcholine*. The most powerful natural anticholinergics are *hyoscine* and *atropine*.

Arsenic (As) is element number 33, atomic weight 75. The element exists in two forms: grey arsenic, which is dense like a metal, and yellow arsenic, which is the non-metallic form. Arsenic reacts with oxygen to form white arsenic trioxide (As_2O_3), often referred to simply as the poison, arsenic.

Arsine is arsenic trihydride (AsH_3). It is a highly toxic gas.

Atomic absorption spectroscopy (AAS) detects the amount of an element in a sample by vaporising it in a hot flame, or by using a laser. The wavelength of the light emitted by the excited atoms identifies the element, and the intensity of the light allows the quantity present to be calculated. The method is quick and has the capability to measure nanogram quantities.

Atropine when pure consists of white crystals which melt at 114 °C. It is a poisonous, bitter, crystalline alkaloid obtained from belladonna and related plants.

Benzene has the formula C_6H_6; its ring structure confers stability to the molecule.

Bitrex is a non-toxic white powder which is soluble in water, and is listed in the *Guinness Book of Records* as the bitterest substance known. Bitrex can be detected at only 2 ppm in water and its taste lingers in the mouth for a long time. Its chemical name is denatonium benzoate, and it is the denatonium portion which is the active component.

The **blood–brain barrier** protects the brain from alien chemicals which find their way into the blood stream and which would seriously interfere with the way the brain operates. It consists of a special lining of tightly knit cells along the 400 miles of blood vessels which supply the brain with essential nutrients such as glucose and oxygen. To penetrate the barrier, molecules have to be small and fat-soluble. Simple molecules such as alcohol, *caffeine, nicotine,* cocaine and *heroin* are able to do so. Although glucose is not fat-soluble, the blood–brain barrier has special entry ports which recognise this molecule and certain other nutrient molecules, and allows them to pass.

Body mass index (BMI) is calculated by taking a person's weight in kilograms and dividing it by the square of their height in metres. It is independent of gender. (It can alternatively be calculated by dividing a person's weight in pounds by the square of their height in inches and multiplying the answer by 700.) Ideally the BMI should lie between 20 and 25. Below 20, a person is underweight. Above 25 they are overweight, above 30 they are obese, and above 40 they are described as morbidly obese, which means their health is compromised by their condition.

Caffeine is found in tea leaves, coffee beans, cocoa beans, maté leaves and cola nuts. Theophylline has a similar structure but without the methyl group attached to the nitrogen of the five-membered ring.

Capillary electrophoresis separates the chemicals in a sample by dissolving them and allowing them to move through a glass capillary tube under the influence of an electric field.

Carbonyl chloride ($COCl_2$) is made by reacting carbon monoxide (CO) with chlorine gas: $CO + Cl_2 \rightarrow COCl_2$. It is the raw material for making polycarbonate. This tough transparent polymer is used for the roofs of conservatories, vehicle lights, CDs and DVDs, and feeding bottles for babies.

Chelating agents are chemicals which can bind to a metal *via* two or more atoms which grip the target metal atom rather like the claws of a crab. (The word 'chelate' is derived from the Greek for a crab's claw.) Chelating agents are used in medical treatments to remove unwanted metals from the body. A particularly powerful example is **EDTA**.

Chirality is a property of certain molecules which arises from their three-dimensional structure. Molecules may have the same atoms arranged in exactly the same molecular form and yet be different by virtue of chirality, in other words they are mirror images of each other. The older term for this property was optical activity, because such pairs of molecules could be distinguished by the way they rotated a beam of polarised light. One molecule would twist such a band of light to the left and be labelled as the D, or *d*, form (short for *dextro*) and the other would twist it to the right and be labelled L, or *l*, (short for *laevo*). These are the Latin terms for right and left, and are still in use. The prefix DL (or *dl*) signifies a 50 : 50 mixture, known as a racemic mixture, which overall does not rotate the light in either direction. Chemists have also used + to signify *dextro* and − to signify *laevo*, and ± to indicate a racemic mixture. Nowadays there is a more sophisticated way of denoting chirality, relating to the exact structure of the molecule.

Chloroform is trichloromethane ($CHCl_3$), and is a liquid (boiling point 62 °C) with a heavy sweetish vapour. It is sensitive to light, which slowly decomposes it, but the addition of 1% alcohol acts as a stabiliser. Although chloroform is only slightly soluble in water, to the extent of 5 ml per litre, it is miscible with most other solvents. Chloroform is produced by heating together chlorine (Cl_2) and methane (CH_4), and is itself used as a solvent. An earlier method of making chloroform was to heat together alcohol and bleaching powder (calcium hypochlorite, $Ca(OCl)_2$).

Chromatography is the most commonly used method of separating a mixture into its component parts and can be effective with quite tiny amounts. Various techniques are available, such as paper chromatography in which a liquid mixture is separated on absorbent paper; thin-layer chromatography, where the liquid moves along a layer of an inert medium such as alumina spread on a glass plate; gas chromatography, in which a volatilised mixture flows over an inert solid coated with a non-volatile fluid; or high performance liquid chromatography (HPLC), in which the sample is forced through a chromatography column under pressure. The chromatographic process can be linked to *mass spectroscopy*, in order to identify the individual components after they have been separated.

Cimetidine is used as an anti-ulcer treatment.

Citalopram is an organofluorine compound with anti-depressant properties.

Codeine is a derivative of *morphine* in which one of the HO groups has been converted to a methoxy (CH_3O) group. It is also known as 3-methylmorphine.

Cyanide is the negative ion CN^- in salts such as sodium cyanide; see also ***hydrogen cyanide***.

Cypermethrin is a powerful insecticide.

Cyperquat is one of a range of ***quaternary ammonium*** herbicides – see ***paraquat***. It is structurally similar to paraquat but instead of two pyridine rings it has one pyridine and one benzene ring (see below). Its full name is 1-methyl-4-phenyl-pyridine (MPP^+). A derivative of Cyperquat is 1-methyl-4-phenyl-1,2,3,6-tetrapyridine ($MPTP^+$), which was inadvertently made by designer drug experimenters in California and caused them to be affected by Parkinson's disease, to the extent that this drug is now used in researching this condition.

Diamorphine is more correctly called diacetylmorphine, or 3,6-diacetylmorphine, and is popularly known as heroin. It is a derivative of morphine in which the HO groups have been converted to acetoxy groups, $CH_3C(O)O$, thereby facilitating the molecule's ability to pass the ***blood–brain barrier***.

1,2-Dichloroethane has the chemical formula $ClCH_2CH_2Cl$ and was formerly known as ethylene dichloride. In even earlier times it was called chloric ether.

Diethamquat is a herbicide with a *quaternary ammonium* structure. It is similar to *paraquat* but with diethylaminocarboxy groups attached:

Diethyl ether is a colourless volatile liquid, with boiling point only slightly above room temperature (35 °C) and its vapour is much denser than air. It is also known as ethyl ether, or just ether. Its formula is $CH_3CH_2OCH_2CH_3$.

Difenzoquat is a *quaternary ammonium* herbicide (trade name, Avenge) in which only one of the nitrogen atoms is in the quaternary state. The positioning of the nitrogens with respect to each other overcomes the problem associated with *paraquat*. The counterion is methyl sulfate. Difenzoquat is particularly effective against wild oats present in crops such as alfalfa, wheat and barley.

4-Dimethylaminophenol (4-DMAP) is an antidote to cyanide poisoning.

Diquat is a *quaternary ammonium* herbicide used as the dibromide. It is available in many countries as Reglone.

Double-blind tests are conducted under conditions where neither the recipient nor the person handing out the drug knows whether the medication is the test material or merely a placebo.

EDTA is an abbreviation for ethylenediaminetetraacetic acid. Sodium and calcium salts of this complex acid make excellent *chelating agents*, especially useful for solubilising and holding on to metals, thereby facilitating their removal from the body.

An **electron microscope** uses a beam of electrons in place of light for examining specimens. Due to their shorter wavelength, electron beams enable much smaller objects to be observed in detail.

Endothelin is a polypeptide consisting of 21 amino acid units and is involved in maintaining optimum blood pressure. The body needs a balance between molecules which tend to restrict the blood vessels, such as endothelin, and others, like nitric oxide, which tend to relax them. Sometimes the body overproduces endothelin, causing blood pressure to rise.

EPA is short for the US Environmental Protection Agency.

Epinephrine is chemically 4-[1-hydroxy-2-(methylamino)ethyl] benzene-1,2-diol. In its pure form it is colourless, but on exposure to light the crystals gradually become brown. It is not very soluble in water but becomes so if acid is added.

Ethanoic acid has the formula CH_3CO_2H.

Ethene, also known as ethylene, has the formula $H_2C{=}CH_2$.

Ether – see *diethyl ether*.

FDA is short for the US Food and Drugs Administration.

Fluoroacetate, also known as fluoroethanoate, is the ethanoate ion $(CH_3CO_2^-)$ in which one hydrogen atom is replaced by fluorine

($CFH_2CO_2^-$). It is a deadly poison which occurs naturally in some African plants.

Formalin is a 40% aqueous solution of formaldehyde, H_2CO, used for preserving anatomical specimens.

Gas chromatography – see *chromatography*.

Half-life is the time taken for half the atoms in a sample of radioactive material to decay. For example, if there are a million atoms present and the half-life is 10 days, then after this time there will be 500 000 atoms remaining. After a further 10 days there will be 250 000, and after another 10 days there will be 125 000, until after 100 days (*i.e.* a total of ten half-lives) only one atom remains. When applied to alien substances in the human body the term half-life refers to the time taken for the excretory mechanisms of the body to reduce the amount by half.

Heroin – see *diamorphine*.

High performance liquid chromatography – see *chromatography*.

HPLC – see *chromatography*.

Hydrogen cyanide (HCN) is a liquid which boils at 26 °C, and can therefore be a gas at room temperature. It is produced by the reaction of methane with ammonia in the presence of oxygen at about 1200 °C over a platinum catalyst:

$$CH_4 + NH_3 + 1.5O_2 \rightarrow HCN + 3H_2O$$

In the laboratory small amounts of HCN may be produced by the addition of acids to the cyanide salts of alkali metals, for example:

$$H^+ + NaCN \rightarrow HCN + Na^+$$

Hydromorphone is *morphine* which has been hydrogenated to add two hydrogen atoms to the non-benzene double bond. It is also known as Dilaudid.

Hydroxycobalamine is a form of vitamin B_{12} in which the cyanide group has been replaced by a hydroxy (OH) group. It is also described as vitamin B_{12b}.

Hyoscine differs from *hyoscyamine* in that it contains an additional oxygen atom, forming a three-membered ring.

Hyoscyamine as produced in nature is the left-handed molecule written as L-hyoscyamine. It is converted to DL-*atropine* by heating with an alkali, and this conversion invariably occurs while it is extracted from natural sources. (See also *chirality*.)

Infrared spectroscropy can identify a material by passing a beam of infrared light through it (or by reflecting a beam of infrared light off it), and comparing the pattern of the absorbed (or the reflected) light with that of a known sample.

Injections can be subcutaneous, which means they are injected under the skin but not into a blood vessel, or intramuscular, which means they are injected into a muscle, or intravenous, which means they are injected directly into a vein.

Kelocyanor is dicobalt-*EDTA* and is an antidote for cyanide poisoning. In this case it is the cobalt atom which reacts with the cyanide ion.

LD$_{50}$ is the lethal dose which kills 50% of a test population, generally assessed using mice or rats. This test was at one time legally required for all potentially dangerous products. A version of it is still a prerequisite for registering a new chemical before it goes on the market.

Lead acetate is more correctly known as lead(II) ethanoate, $Pb(CH_3CO_2)_2$. It is readily soluble in water and the solution has a sweet taste.

Limestone is calcium carbonate ($CaCO_3$), and when heated in a kiln loses carbon dioxide (CO_2) to form *quicklime* (CaO). The latter has many uses, such as making plaster casts or for reducing the acidity of soil. It reacts vigorously with water, giving off heat as it does so, and forms slaked lime, calcium hydroxide, $Ca(OH)_2$.

Linamarin is a cyanide occurring naturally in cassava, lima beans and flax.

Lissapol N is the nonionic surfactant, poly(ethyleneglycol)nonyl-phenyl ether. It has the formula $H(OCH_2CH_2)_nOC_6H_4C_9H_{19}$ and the molecular structure:

Mass spectroscopy (MS) is an analytical technique in which a beam of molecules is ionised, thereby breaking them into various smaller ions. These are then passed through a strong electric and a magnetic field which separates them according to their mass and charge and allows them to be identified. This technique is particularly useful in differentiating isotopes of the same element.

Methionine is an essential amino acid, which means it has to be part of the diet because it cannot be made within the human body. It can be produced synthetically by reacting methyl mercaptan (CH_3SH) and acrolein ($CH_2\!=\!CHCHO$) with HCN, CO_2 and NH_3.

Morphine has the molecular composition $C_{17}H_{19}NO_3$. Its mono-hydrate ($C_{17}H_{19}NO_3.H_2O$) is crystalline but sparingly water-soluble (only 0.2 g per litre). However, it is more readily soluble as the sulfate or the hydrochloride, which have solubilities around 60 g per litre. Related substances include *codeine, diamorphine (heroin), hydromorphone*, and *naloxone*. The molecular structure of morphine contains five *chiral* centres, and it was this factor which made its laboratory synthesis so difficult when first attempted. It was eventually achieved in 1955 by Marshall Gates and Gilg Tschudi at Rochester University in the USA. The opium poppy uses natural enzymes in its manufacture, and manages this so readily that it is the only economic source of morphine.

Naloxone is a derivative of morphine.

Nuclear magnetic resonance spectroscopy (NMR) is used to identify the location of atoms in a molecule by the radio frequencies which they absorb in a powerful magnetic field. This causes the nuclei of atoms such as hydrogen and carbon–13 to change their direction of spin. The energy required to do this is influenced by the electrons surrounding the nucleus and is characteristic of the chemical bonds present in the molecule.

Neutron activation analysis (NAA) is an analytical technique which is so sensitive that it can measure the amounts of certain metallic elements in a single strand of hair. The drawback is that it requires access to a nuclear reactor. When atoms are exposed to neutrons inside such a reactor they are absorbed and can be converted to specific short-lived isotopes which decay, emitting characteristic radiation such as gamma rays. Measurement of the radiation allows identification of the atoms present. The technique is sufficiently sensitive to enable quantities in the nano-gram or even picogram range to be assessed.

Nicotine is 3-(1-methyl-2-pyrrolidinyl)pyridine and occurs natu-rally in the leaves of *Nicotinia tabacum* to the extent of 8% (dry weight). It can be used as an insecticide and is highly toxic to humans, although in small doses it relaxes the brain. It is ad-dictive, however.

Nylon is formed by condensation polymerisation (which means that water is eliminated during the process) of 6-aminohexanoic acid. The product is known as nylon 6. An analogous polymer is formed by reaction between hexanedioic acid (adipic acid) and 1,6-diaminohexane, in which case the product is described as nylon 6,6 and has the following structure:

Optical isomers – see *chirality*.

Organophosphorus compounds are those containing a covalent bond between phosphorus and carbon, whereas organophosphates are organic salts of phosphoric acid, H_3PO_4. DNA is thus an organophosphate, as are a number of common insecticides. Organophosphorus compounds tend to be deadlier in their effect and include, for example, the nerve gases.

Paraquat is 1,1'-dimethyl-4,4'-bipyridinium and is generally available as the dichloride salt. It is a *quaternary ammonium* compound with powerful herbicidal properties. It is highly toxic due to the fact that the cells of the body mistake it for a peptide fragment on account of the relative positions of its two positively charged nitrogen atoms with their methyl groups. By changing the locations of these nitrogens, or by removing one of them, or by attaching other groups to them, it has been possible to make safer herbicides such as *cyperquat, diquat, diethamquat* and *difenzoquat*. Paraquat is also known as methyl viologen. In its reduced form paraquat, having acquired an extra electron, turns bright blue: this makes it suitable for use as an indicator in redox reactions.

Pholcodine is a derivative of morphine in which one of the HO groups has a bulky morpholinoethyl group attached:

Phosphorus (P) in its common elemental form is a waxy white solid of composition P_4. It phosphoresces in the dark, is spontaneously flammable, and is highly poisonous. Another form is red phosphorus (P_n), which possesses none of these features.

Polonium (Po) is an element with atomic number 84, and all its isotopes are radioactive. It is a chemically reactive silver-grey semi-metal which dissolves in dilute acids. The solution is initially pink, due to the Po^{2+} ion, but rapidly turns yellow since the α-radiation present forms oxidising species with water molecules, and these convert it to Po^{4+}.

Polyester is a polymer formed by the reaction of a molecule with two hydroxy groups reacting with an acid containing two carboxylic acid groups. The most common commercial polyester is polyethylene terephthalate (PET), which is formed from ethylene glycol (1,2-ethanediol) and terephthalic acid (1,4-benzenedicarboxylic acid).

Poly(methylmethacrylate) is chemically poly(methyl 2-methylpropenoate). It is a polymer noted for its strength and transparency. Its trade names include Perspex in the UK and Lucite in the USA.

Poly(propenonitrile) is known as polyacrylonitrile (PAN) and marketed under trade names such as Acrilan and Courtelle. Fibres made from it are noted for their softness.

ppb is an abbreviation for parts per billion, and is therefore 1 µg per kg or 1 µg per litre. See also **units**.

ppm is an abbreviation for parts per million, *i.e.* 1 mg per kg, or per litre. See also **units**.

Propanone, better known by its familiar name, acetone, is a volatile liquid used as a solvent and has the formula $(CH_3)_2CO$.

Prussian Blue is chemically ferric ferrocyanide, $Fe_4[Fe(CN)_6]_3$. There are a number of pigments of similar chemical structure, including Berlin Blue, Chinese Blue, Bronze Blue, Non-bronze Blue and Milori Blue. As a printing ink pigment it has high tinting strength but only moderate light fastness. It is used either in its own right, particularly as Bronze Blue, which has a characteristic reddish sheen, or in black inks as a means of counteracting brownness. The analogous compound ferrous ferricyanide, $Fe_3[Fe(CN_6)]_2$, is a pigment known as Turnbull's Blue.

Quaternary ammonium refers to a nitrogen atom bonded to four other atoms and which therefore carries a positive charge. *Paraquat* is a *quaternary ammonium* herbicide with two such nitrogen atoms within its molecular structure.

Quicklime is the common name for calcium oxide (CaO). It is produced by heating **limestone** (calcium carbonate, $CaCO_3$) at high temperature, when it loses carbon dioxide. Quicklime is colloquially referred to simply as lime, and was traditionally produced in lime kilns to be spread on farmland to make the soil less acidic. It reacts vigorously with water, giving off heat, and forming slaked lime (calcium hydroxide, $Ca(OH)_2$).

Ribosomes are tiny particles present in large numbers in all living cells, and where proteins such as enzymes are made. Ribosomes are about $2\,\mu m$ in diameter and consist of about 40% protein and 60% RNA. Ribosomes are the sites at which information carried in the genetic code is applied to make protein molecules. The newly formed proteins detach themselves from the ribosome site and migrate to be used in other parts of the cell.

Ricin consists of two long chains, described as A and B, which are connected by a disulfide bridge (–S–S–). Chain A is the deadly component and contains 267 amino acids and a molecular weight of around 30 000 daltons. Chain B has 262 amino acids and a similar molecular weight. Chain B has the ability to bind to a specific carbohydrate unit, a galactose group, on the outside of cells. Once attached it waits for chain A to pass through the cell membrane. Once inside the cell, chain A seeks out the *ribosome* that makes the essential enzymes the cell needs and blocks this by targeting a specific base unit, *adenine*, in the RNA. Adenine has a double ring structure and this becomes entangled with part of chain A, putting all the RNA out of action. Without its supply of essential enzymes the cell then dies.

Ricin can be extracted from castor beans, but the process is hazardous and can put life at risk. Ricin can only be safely handled in a specially equipped laboratory.

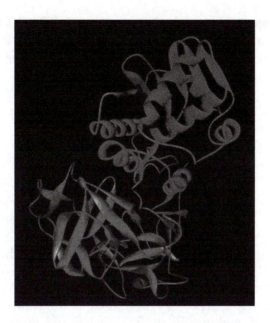

Ricinoleic acid is the best known of the straight-chain hydroxy fatty acids, and has the formula shown. It occurs in mature castor plants and is known as castor oil acid. When heated with sodium hydroxide it cleaves to form sebacic acid, $HO_2C(CH_2)_8CO_2H$, which is a raw material for the chemical industry.

Silica is the common name for silicon(IV) oxide (SiO_2) and occurs naturally as quartz, flint, sand, *etc.* It is extremely stable chemically and can be used to line furnaces. When fused with sodium carbonate and calcium oxide it forms ordinary glass.

Silicones, or more correctly siloxanes, are polymers consisting of long chains (or rings) of alternate silicon and oxygen atoms with two organic groups attached to each silicon atom. In most silicones these are methyl groups. Silicones are highly water-repellent and widely used in waterproofing treatments and sealants.

Sodium nitroprusside has the formula $Na_2[Fe(CN)_5NO]$ and forms ruby-red crystals which can provide a test for aldehydes and ketones. It has also been prescribed for reducing high blood pressure.

Superoxide is an oxygen molecule, O_2, which has picked up a free electron to become an $O_2\cdot$ radical, which is a powerful oxidising agent.

Syngas is obtained when steam is blown through red hot coke, coal or charcoal, or when natural gas and steam are heated together. In the presence of an iron or cobalt catalyst syngas can be converted to hydrocarbons, especially those used as transport fuels.

This way of making liquid fuels from coal was developed in countries which found themselves forced to rely on their own resources for energy, for financial reasons, war or global boycotts. Syngas was made by a process devised by the Lurgi Company of Germany and is converted to oil by the Fischer–Tropsch process developed in the 1920s. South Africa still employs Lurgi/Fischer–Tropsch plants to meets its energy needs and Sasol, the company operating them, turns 30 million tonnes of coal a year into transport fuels, including low-sulfur diesel, and into chemicals such as plastics.

Tartaric acid can exist in three chiral forms known as *meso-*, D- and L-tartaric acid – see ***chirality***. Tartaric acid is a natural product which crystallises in the form of the potassium salt from vats of wine. It is used in baking powders.

Tartar emetic is the common name for antimony potassium tartrate – see ***tartaric acid***. The antimony potassium salt was prescribed by doctors in the 1700s and 1800s to reduce the temperature in fevers, and sometimes to make people vomit—hence its common name.

Tetrachlorordibenzodioxin, also known as TCDD, is the most toxic of the dioxins. Dioxins contain two benzene rings with one or more chlorine atoms attached, and in the case of TCDD there are four of these, in positions 2, 3, 7 and 8.

Theophylline – see *caffeine*.

Thin layer chromatography – see *chromatography*.

Units for discussing toxicological data need to encompass tiny amounts and concentrations, such as milligrams (mg) per litre and micrograms (µg) per litre. Alternatively, these quantities can be expressed as parts per million (*ppm*) or parts per billion (*ppb*), respectively. To give an idea of the size of this last quantity, 1 ppb can be compared with 1 second in 30 years. It is even possible for modern analytical techniques to measure micrograms per millilitre, which corresponds to parts per trillion (ppt); this is like 1 second in 30 000 years. In terms of weight, 1 ppm is of the order of a grain of sugar in a one-kilogram bag, and 1 ppb is on the scale of one grain in a tonne of sugar.

Verapamil has been used in the treatment of angina.

X-Ray fluorescence spectroscopy (XRF) is an analytical technique using high-energy radiation from a radioactive isotope such as promethium–147, which knocks out an electron from the innermost orbit of an atom. This electron is then replaced by an electron from the next orbit, and as it does so it emits an X-ray characteristic of that particular element.

Zeolites are aluminosilicates, and can be either natural or synthetic. They make excellent catalysts because they contain large cavities into which smaller molecules can diffuse and be brought into close contact, making it easier for them to react.

Metric/Imperial Conversion Tables

Imperial/USA unit	Metric (SI) unit
Inch	2.54 centimeters
Foot	30.48 centimeters
Yard	0.91 meters
Mile	1.61 kilometers

Metric (SI) unit	Imperial/USA unit
Centimeter	0.39 inches
Meter	3.28 feet
Meter	1.09 yards
Kilometer	0.62 miles

Imperial/USA unit	Metric (SI) unit
Ounce (weight)	28.35 grams
Pound	0.45 kilograms
UK ton (2240 pounds)	1.02 metric tons
US ton (2000 pounds)	0.91 metric tons

Metric (SI) unit	Imperial/USA unit
Gram	0.035 ounces
Kilogram	2.21 pounds
Metric ton (1000 kg.)	0.98 UK tons
Metric ton (1000 kg.)	1.10 US tons

Subject Index